# China's Grain for Green Program

Claudio O. Delang • Zhen Yuan

# China's Grain for Green Program

## A Review of the Largest Ecological Restoration and Rural Development Program in the World

Springer

Claudio O. Delang
Zhen Yuan
Department of Geography
Hong Kong Baptist University
Kowloon Tong, Hong Kong

ISBN 978-3-319-11504-7 ISBN 978-3-319-11505-4 (eBook)
DOI 10.1007/978-3-319-11505-4
Springer Cham Heidelberg New York Dordrecht London

Library of Congress Control Number: 2014950668



Printed on acid-free paper

Springer is part of Springer Science+Business Media (www.springer.com)

# Preface

China has always had a problem of inadequate supply, low quality, and uneven geographic distribution of forest resources. However, the second half of the twentieth century saw the greatest deforestation in the country's history, as the rapid increase in the nation's population, coupled with rapid economic development, resulted in enormous consumption of forest resources. Forest coverage decreased from 30–40 % in 1949 to about 10 % in the late 1990s. By the end of the twentieth century, while China's population accounted for 22 % of the world's population, the forest area in China was only taking up 4.1 % of the world's land mass, and the stocking volume was merely 2.9 % of the world's total (Lei 2002). This was clearly not sufficient to meet the production and livelihood needs of the country. However, even more pressing were the environmental problems that decades of deforestation had created.

Excessive commercial logging, and the cutting down of the forest on hillsides for cultivation in the upper and middle reaches of the river basins, led to severe consequences in downstream areas. Using data from China's second soil census of the early 1980s, Yang (1994) found that around 8 % of the country's cultivated land was affected by "intensive" erosion, and another 26 % was affected by "light to medium" erosion. By the turn of the millennia, 170 million ha, 18.2 % of the country's land, were desertified, affecting 400 million people. On the other hand, 360 million ha of soil nation-wide were eroded, which accounted for 38.2 % of the country's total land area, or more than three times the world average (Lei 2002; Huang 2000). Soil losses reached 5 billion tonnes annually (Lei and Zhu 2002; World Wildlife Fund and State Forest Administration 2003).

The situation was particular dire in the watersheds of the Yangtze and the Yellow rivers. The main channels of Yangtze River flow through 11 provinces in China. The size of the river basin is roughly 1.8 million $km^2$, or 18.75 % of China's land area. The river basin is an extremely important area for the economic and social development of China, since it produces 42 % of China's GDP and hosts 43 % of the country's fixed investment. The Yellow River basin covers over 900,000 $km^2$, and flows through nine provinces. With a total population of some 190 million people, it is not only economically, but also culturally, important, being the birthplace of the ancient Chinese civilization.

According to official estimates, at the turn of the century the soil erosion area in the Yangtze and Yellow river basins reached 75 million ha, with sediments of over 2 billion tonnes (Li 2001). Overgrazing, and in particular farming on slopes, were the most important causes of soil erosion and desertification. Xu et al. (2006a) estimated that of the 34.07 million ha of farmland in the Yangtze and Yellow river basins, 4.25 million ha were on slopes of 25° or greater. Farming such slopes was estimated to increase erosion to 4,000 tonnes per $km^2$ per year. With proper forest coverage, 80–90 % of that erosion could have been prevented (Yin et al. 2005). On the other hand, the Loess Plateau (containing the upper watershed of the Yellow River) was estimated to contain 22 % of China's eroded land, and 19 % of China's cultivated area affected by "intensive" water erosion. On the Loess Plateau, uncontrolled grazing and poor maintenance of rangelands were the main causes of the extensive loss of grass cover, and contributed to soil erosion.

Increased soil erosion silted streams, reduced the hydraulic capacity of the rivers, and increased the frequencies of flooding and drought (Smil 1993; World Wildlife Fund and State Forest Administration 2003). While during historic times there were regular flooding disasters in the lower reaches of the Yellow River, towards the end of the twentieth century the situation reversed. During the dry season, the water flow sometimes ceased in parts of the lower reaches. This happened for the first time in 1960. After 1972 it happened frequently, and since 1992 it has been happening every year. In 1997, there was no water discharged to the sea for 330 days (Fu et al. 2004), and the water flow was interrupted for up to 700 km upstream from the river mouth. The seasonal interruption of the water flow cannot only be blamed on deforestation. The diversion of water for urban and farmland water supply also takes some blame. However, deforestation upstream contributed to the problem (Wang et al. 2001).

From June to September 1998 there were devastating floods in the middle reaches of the Yangtze River. The flood affected 180 million people, and resulted in some 4,000 death and 15 million homeless. A total of 13.3 million houses were damaged or destroyed, and 10 million ha were evacuated. The economic losses accounted for some US$26 billion. Many environmental experts blamed these floods on soil erosion and deforestation (World Bank 2001). For example, Zong and Chen (2000) argue that the amount of precipitation over the catchment and the floodwater discharge from the upper basin did not exceed the historical maximum. Rather than being caused by the increasing precipitation, downstream floods were caused by extensive reclamation of lakes and fluvial islands, deforestation in the catchment area, and soil erosion and the resulting increasing deposit of sediments in reservoirs, which reduced their storage capacity.

By the late 1990s, it became obvious that China was facing serious environmental problems caused by decades of mismanagement of forest resources. At the same time, the Chinese government estimated that over the next 50 years the demand for forest resources in China would reach at least 18.5 billion cubic meters, 1.6 times the forest resources that existed in the late 1990s (Lei 2002). These problems prompted the government to act, and in the late 1990s the Chinese government introduced a number of reforestation and ecological restoration programs, the most important of which was the Grain for Green (GfG) (also called Slope Land Conversion Program, SLCP).

The GfG was put in place primarily to reconvert steep slopes that had been cleared for farming to their original vegetation (trees or grassland), thereby reducing siltation in the rivers. As the prime managers of the reforestation processes, the farmers would be compensated for their labor and loss of agricultural land. Therefore, the GfG is not only a reforestation and ecological restoration program, but also a poverty alleviation program. The GfG started in 1999 in three selected provinces, and expanded nation-wide starting from 2000.

Bennett (2008) and Wu et al. (2009) reported that the program planned to convert a total of 32 million ha of land to its original vegetation (trees or grass) during the period from 2001 to 2010: 14.66 million ha of farmland (4.4 million of which was estimated to be on land with slopes of 25° or above), and 17.33 million ha of barren mountainous wasteland. By 2010, the forest and grass cover of the scheme's target area would be raised by 5 %; 86.66 million ha of soil- and water-eroded area would be brought under control, and 103 million ha of sand-fixation areas would be established (Lei 2002).

The GfG is the reforestation, ecological restoration, and rural development program with the largest investment, greatest involvement, and broadest degree of public participation in history. The program improves the ecological conditions of much of China, and the socioeconomic circumstances of hundreds of millions of people. The GfG directly involves 124 million people (32 million households) in 1,897 counties in 25 provinces and the Xinjiang Production and Construction Corps (Mao et al. 2013). The program was set to end 8 years after it initially began, but was extended for another 8 years in 2007. It is now set to end starting in 2015 (later in areas where it started later). By the end of 2015, the government is expected to have invested no less than Yuan 431.8 billion (National Development and Reform Commission 2008).

This book reviews the literature pertaining to, or related to, the GfG, published up to 2014. The book is organized in three parts. Part I introduces the conditions that led to the introduction of the GfG, and compares the GfG to the other main reforestation programs in China. Part II gives an overview of the GfG, describing the timeline of the program, the compensation paid to farmers, the rules concerning land and plant selection, and the extent to which these rules were followed. It also discusses the attitudes of farmers towards the program, and the way in which the program is organized and implemented by various state actors. Part III discusses the impact of the GfG, both from an ecological and from a socio-economic standpoint. The focus is on the socio-economic consequences of the program, and in particular the economic benefits that result from participating in the GfG, the impact of the GfG on the local economies, and the redistribution of the labor force. We also consider the sustainability of the program, since the question arises as to what will happen to the converted land when payments to farmers end.

Kowloon Tong, Hong Kong

Claudio O. Delang
Zhen Yuan

# Contents

## Part II Overview of the Grain for Green

# List of Figures

# List of Tables

# Part I
# Why the Grain for Green?

The first part of the book looks at the conditions that prompted the government to introduce the Grain for Green program in 1999. This part is divided into two chapters. The first chapter discusses the forest policies from 1949 to 1998. In China forests were (and often still are) seen largely as uncultivated farmland. Until 1949 there was not even a Ministry of Forestry, with the forests being managed by the Ministry of Agriculture. The communist government did try to use forests more rationally, by instituting the first national-level Ministry of Forestry (MOF). However, the MOF was not able to manage the forests sustainably, being understaffed and with insufficient funds to replant the trees that had been cut. The objectives of forest management were basically to help promote the development of the country: from 1949 to 1998, the role of timber was to produce cheap raw material and fuel for the national drive towards industrialisation, and the role of forestland was to provide agricultural land to feed the burgeoning population. In particular the Great Leap Forward resulted in great deforestation, as farmers cut large amounts of timber to aliment furnaces to melt pig iron, and as forests were cut to increase the amount of farmland. Deforestation also continued during the Cultural Revolution, when the government (including the MOF) was scaled down, much of its staff removed, and the feeble attempts to control deforestation were further weakened. After 1978 there have been a number of reforms, including the Resolution on Issues Concerning Forest Protection and Development in March 1981, popularly known as the "Three Fixes", and the Forest Law of 1984. These helped improve the situation, but did not stop deforestation from continuing.

The very extensive deforestation that occurred since 1949 culminated in the drought of the Yellow River in 1997, and the flooding of the Yangtze River in 1998. The Yellow River had been drying up during the summer for years, but in 1997 the river dry-up period lasted 227 days at the Lijin Hydrological Station (100 km upstream from the river mouth), and for 330 days there was no water discharged into the sea. This put industrial, agricultural and residential water uses in great jeopardy. On the other hand, in 1998 there were massive floods along the Yangtze River, which claimed the lives of some 4,000 people, displaced 18 million people and led to more than US$12 billion in property damage and output loss. These calamities led the government to introduce the reforestation and environmental conservation programs discussed in Chap. 2.

Chapter 2 describes the six largest programs introduced during the late 1990s and early 2000s. Together, these six Key Forestry Programs (KFPs) cover 97 % of China's counties and target over 100 million ha of land for forestation. The GfG is the largest of these programs, in terms of area covered, people affected, and money invested. This book only reviews the GfG, but in many villages more than one program was introduced concurrently. We argue that the government introduced the KFPs not only because environmental deterioration had reached a critical point, but also because China was producing a surplus of grain, which lowered farmers' incomes, and because inequality between the eastern and western provinces was reaching a critical point. The GfG in particular addressed all these problems concurrently, through direct payments to poor farmers willing to set aside marginal land. Partly for this reason, the GfG is considered by many as the best reforestation and rural development program ever undertaken in China.

# Chapter 1
# Management of Forest Resources from 1949 to 1998

**Abstract** The first chapter discusses the forest policies from 1949 to 1998. In China forests were (and often still are) seen largely as uncultivated farmland. Until 1949 there was not even a Ministry of Forestry, with the forests being managed by the Ministry of Agriculture. The communist government did try to use forests more rationally, by instituting the first national-level Ministry of Forestry (MOF). However, the MOF was not able to manage the forests sustainably, being understaffed and with insufficient funds to replant the trees that had been cut. The objectives of forest management were basically to help promote the development of the country: from 1949 to 1998, the role of timber was to produce cheap raw material and fuel for the national drive towards industrialization, and the role of forestland was to provide agricultural land to feed the burgeoning population. In particular the Great Leap Forward resulted in great deforestation, as farmers cut large amounts of timber to aliment furnaces to melt pig iron, and as forests were cut to increase the amount of farmland. Deforestation also continued during the Cultural Revolution, when the government (including the MOF) was scaled down, much of its staff removed, and the feeble attempts to control deforestation were further weakened. After 1978 there have been a number of reforms, including the Resolution on Issues Concerning Forest Protection and Development in March 1981, popularly known as the "Three Fixes", and the Forest Law of 1984. These helped improve the situation, but did not stop deforestation from continuing.

**Keywords** Forest policies • Historical deforestation in China • Great leap forward • Cultural revolution • Household Contract Responsibility System (HRS) • Three Fixes

## Introduction

At the time of liberation, there were two pressing problems in China. First, it needed to feed the large and burgeoning population. In 1949, the total grain output was 1,130 million tonnes (CSY 1991), or 209 kg per capita – less than the United Nations' Food and Agriculture Organization (FAO) recommendation of 220 kg of grains per capita for a healthy diet. Second, China was a rural country, and it was



C.O. Delang, Z. Yuan, *China's Grain for Green Program*,
DOI 10.1007/978-3-319-11505-4_1

felt that the nation needed to industrialise as fast as possible, particularly by developing the manufacturing sector. As Lenin (Lenin 1972: 549) put it, "Heavy industry is the only possible economic base of socialism."

The Chinese government began addressing the problem of food scarcity by transforming large areas of forestland into agricultural land, thereby undertaking massive deforestation (Du 2002). Between 1949 and 1979, 38 million hectares (ha) of forestland, wasteland and wetland were transformed into farmland (Feng et al. 2005a). Furthermore, to support industrial production during the Great Leap Forward, huge swaths of forests were transformed into firewood, mainly for the production of pig iron (Li 1985; Du 2002; Richardson 1966). While there were some in the Ministry of Forestry (MOF) who spoke up about the importance of sustainable timber extraction and reforestation, their concerns went largely unheard (Li 1988a; Wang 2000).

These twin problems are what drove China's policies regarding state-owned forests, which were addressed within the communist ideology of its leaders, particularly that of Mao Zedong. Far from a progressive, forward-thinking policy of conservation, these two challenges were addressed by extensively exploiting and misusing China's forests. This chapter describes the policies of the Chinese government that affected the forestry sector during the 50 years from the liberation in 1949–1998. These policies led to deforestation, and the environmental and ecological problems of the late 1990s, which forced the government to change course of action, and institute several very extensive reforestation and rural development programs, including the Grain for Green (GfG) program, in the late 1990s (discussed in the following chapter).

## Historical Land Policies in China

In order to fully appreciate the challenges involved in implementing the GfG program, one must understand the land policies that preceded it. One of the earliest policies after the liberation was the Agrarian Reform Law of June 1950, which was laid down by the Government Administration Council (GAC) of the Central People's Government (CPG)[1] and remained valid until November 1987. Under the Agrarian Reform Law, there were two categories of forestland: state-owned and non-state-owned. These would become the most commonly-used categories, in different formats, for several decades.

State-owned and non-state-owned forestlands were further distinguished based on geographic criteria, particularly by whether they were situated in the north or the south (Fig. 1.1). Since forest cover was dense and the population was sparse in the

[1] The Government Administration Council (GAC) of the Central People's Government (CPG) was the highest state administrative organ in China. It was replaced by the State Council on 28 September 1954.

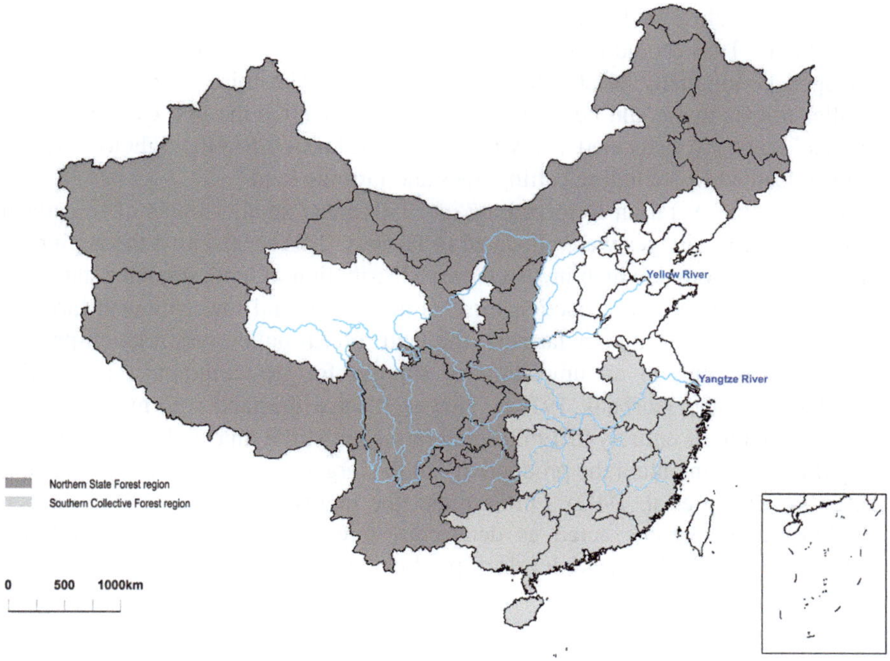


**Fig. 1.1** State forest and collective forest

north, state-owned forest enterprises (called state *linchangs*[2]) were established there. Landless farmers and soldiers were settled on these forestlands, with the goal of producing timber for the national economy.

In the south, population density was greater. Southern forestlands were situated close to human habitations where the communities were better developed, and extensive forest use had been the norm for centuries as part of the livelihood of peasants (Menzies 1988). As a result, most of the forestland in the southern region was non-state-owned, with more than 85 % of the forests in private hands. The government planned to eventually source 46 % of all wood consumed in the country from these nine provinces (FRSOC 1983; Li 1985).

In this chapter, we discuss in more detail the policies targeting the use and misuse of timber and forestland in the southern non-state-owned forests, because this is where the GfG program was implemented. The northern state-owned forests will not be discussed here.

Within the Agrarian Reform Law, national forestland was further divided into three broad categories. The first category included larger tracts of natural forests, which were predominantly located far from villages. These forests were now to be managed directly by the state, but because of their location, they had not been used by people, so there was no loss to the farmers. The second category consisted of

[2] Forest farms, set up by the government to produce forest products and manage the forests.

smaller forests close to villages, which were somewhat difficult for farmers to manage individually. These could be on hilltops or on steep slopes. Such forests were to be managed communally (CLC 1982a [1949–1950]). The third category included smaller forests that could be managed by farmers. Unlike the first category, these 'private' forests were located near villages, and the farmers had the right to use them as they deemed fit, including cutting trees and farming land.[3]

Through the Agrarian Reform Law of 1950, many smaller tracts of forestland located close to villages were allocated to farmers. This represented the beginning of rural land reforms, which involved the redistribution of land, implementation of a fairer tax system, and reduction in rents in order to aid lower-income farmers. Given the benefits to poorer farmers, the land reform policy was a much-praised initiative, resulting in an upwelling of support for the communist authorities (Macfarquhar and Fairbank 1978a). However, once the land reform policy was implemented in its entirety in 1952, small farms with little financial capital found it very difficult to maintain the productivity of the larger pre-reform farms (Du 2002; Ye 2006; Huang et al. 1992). Furthermore, the repressive measures used on rich landlords and peasants acted as deterrents to excessive individual ambitions. Overall, this resulted in lower levels of productivity.

## *From Independent Farms to Communes*

From 1952 onwards, the Chinese Communist Party (CCP) moved to overcome this diminished output by encouraging farmers to join cooperatives. Cooperatives (from 1952 to 1955 they were called 'cooperatives'; from 1956 to 1958 'senior cooperatives', and from the end of 1958 onwards 'communes') were not immediately popular. A small handful of farmers joined at first, but over time as the communist ideology caught on and following official government encouragement to do so, a larger number of farmers eventually organised themselves into cooperatives voluntarily. In 1951, 1,618 households were members of communes, but by 1956, 118 million households (almost the entire Chinese population) were members of communes, with an average of 155 households per commune (Hu 2007; Huang et al. 1992; SAC 1981).

The commune gradually took control of all trees (including those near houses) and non-timber forest products (such as fruits and mushrooms), that originally belonged to its members – sometimes, but not always, with financial compensation for those affected. Before the establishment of communes, farmers could earn a profit from the trees they had planted. After the farmers joined the commune, the trees became part of the commune's assets. Although the farmers were theoretically

[3] The Agrarian Reform Law was often implemented slightly differently in different provinces. For example, in Sichuan province small pieces of forest near or connected to buildings (e.g. farmers' houses, temples, schools) continued to be managed by individual farmers (Wang 1994), while most forestlands larger than 33.3 ha (though in certain counties 20 ha or 6.7 ha) were nationalised.

compensated for the shift in ownership, the compensation they received was much lower than they would have obtained from the sale of the trees. In addition, the compensation was paid incrementally over many years, rather than immediately. Because of the low – and slow – compensation paid to farmers, many of them felled all the trees on their land and sold the timber before joining the communes (Du 2002). Many farmers were also dismayed by the fact that the trees they had been encouraged to plant in the early 1950s were now under the control of communes – so much so that they no longer wanted to plant trees or take care of the forests.

By the end of 1956, 95 % of households had given all their land usufruct (including from private forests) to the communes (Du 2002). All the forestland of the commune members was amalgamated, and a number of workers were assigned to carry out the necessary work. The farmers became essentially government employees who earned a fixed salary by working for the commune; the links between farmers and forestlands were disconnected. Despite severing the connection between farmers and forests, the commune system did provide some benefits to the forest cover, particularly in activities requiring collaborative labor, such as combating forest fires. The frequency of forest fires decreased by 90 % in 1952–1953, compared to 1951–1952 (Li 1988a: 25). The large work parties that could be organised in communes also facilitated reforestation, and 1.1 million ha were reforested in 1953 (Du 2002: 212). However, organising all rural villages into communes had an overall negative effect on the forest cover.

## *The Great Leap Forward*

China adopted the USSR's economic strategies to industrialize the country as rapidly as possible. These strategies aimed for high rates of reinvestment, emphasizing capital-intensive, high-technology projects, and used agriculture as a major source of funding for industrial growth and developing a heavy industry sector represented by the iron and steel industry (Macfarquhar and Fairbank 1978a: 96). The development of the entire country was thought to be represented most significantly by two indicators (out of thousands of possible indicators): steel, which was used to assess the development of the industrial sector; and grain, which was used for the development of the agricultural sector. Hence, the national development plans emphasised the productions of these two products.

In the mid-1950s, the production of steel and grain had surpassed the expectations of Mao Zedong and government officials. Within 1 year, the people achieved a growth in output that the authorities believed would have required 10–20 years. Encouraged by this success, and based on almost utopian optimism,[4] the government

[4] 'Your determination determines your productivity' (in Chinese "ren you duo da dan, di you duo da chan") was a slogan coined during this time for agricultural production, published on the Red Flag, an important official magazine in China during those years (Tao 1958).

set very high standards (Shen 2008). In 1958, Tao Chu[5] argued that since nature has endless potential, farmers could aim at a yield of 5,000 kg of rice per mu (15 mu = 1 ha) (Tao 1958). Since much of the population was living in commune, and following the communist principles of self-reliance, the drive towards growth was centered in these communes. Hence, the government encouraged communes to construct their own small factories to produce what they required to become independent economic units, from pig iron to furniture, and from paper to cooking utensils. By June 1959 (only two-and-a-half years after the Great Leap Forward officially commenced), there were about 700,000 commune enterprises, with an industrial output valued at about Yuan 710 million, equivalent to 10 % of the national industrial output (Yu 1991).

However, small-scale development of manufacturing (especially pig iron production) in the communes required large amounts of wood to fuel the inefficient furnaces. Also, expected increases in agricultural output were unrealistic, and could not be produced on existing land, and so farmers had no choice but to expand the amount of farmland. Unsurprisingly, the result of the Great Leap Forward was vast large-scale deforestation. Indeed, the push for industrial and agricultural expansion impacted the health of forests so much so that the Great Leap Forward became the first period in Communist China history where severe deforestation took place.

While we have no clear data regarding the amount of timber used in steel production nationwide, there are many descriptions confirming extensive deforestation at that time (Richardson 1966, 1990; Tao 1994; Li 1985; Du 2002). In 1958, the government issued a policy requiring the people to afforest all the barren hills by 1970 and increase the forest cover rate above 20 % (CLC 1982b [1957]). However, reforestation was considered less important than increasing the output of steel and grain, so the target was never achieved. A popular saying during these years was that "Everything should service rice and steel," and forests were not a central focus of government policies (Macfarquhar and Fairbank 1978a). Therefore, a person caring for forests was regarded as someone who was opposed to government policies.

From 1959 to 1961, China faced a significant famine due in part to the Great Leap Forward, combined with a widespread drought. In response, the Chinese government terminated the Great Leap Forward and instituted a set of reforms in 1961. The central principle of these reforms was "agriculture as the base, industry as the leading factor [with the aim of] readjustment, consolidation, filling out and raising standards, [replacing the previous formula of] greater, faster, better and more economical results" (Macfarquhar and Fairbank 1978a: 339).

Once the drought passed in June 1961, central government published the "18 Articles on Forestry", the main requirement being for communes to return trees to farmers,[6] stating in particular that "the trees of the commune members who plant trees near roads, rivers, in the village and on tombs, belong to those who plant them"

[5] The governor of Guangdong province and member of the CCP's Central Committee.

[6] The official name of the policy, issued 21 June 1961, was "Rules on confirming forest property, protecting forests and developing forestry (draft)" (CFY 1987). Since it was made up of 18 articles, it was also called "18 articles on forestry". The 18 articles on forestry basically addressed the same

(CFY 1987). Yet even with "the 18 Articles of Forestry" and other laws requiring that trees be returned to those who had invested their seedlings and labor, most commune members were still required to give up their possessions to the commune (Huang 2006). As a result, in most places the laws were ignored, or their implementation was interrupted mid-process as a result of the Cultural Revolution (Huang 2006; SFA 1999).

## *The Cultural Revolution*

The disaster of the Great Leap Forward let to the emergence of people within the Central Committee who were against Mao Zedong. Voices such as those of President Liu Shaoqi and the Party General Secretary Deng Xiaoping began to be heard more loudly as they argued for economic reforms in contrast with those of Mao's communitarian vision. Mao Zedong's response was to launch the Great Proletarian Cultural Revolution on 16 May 1966 to prevent the restoration of capitalism and quieting elements he perceived to be bourgeois, both within the government and within the country as a whole. Political purging resulted in government departments being substantially down-sized, including a widespread reduction of personnel at all levels of the Ministry of Forestry (MOF). Many employees were either demoted or sent to prison, including those in charge of forest management.[7] As a result, several years passed with very little oversight pertaining to forest health and management. Planners who would have been able to address new problems were discouraged from doing so, fearful of political reprisals. In the words of Mac- Farquhar: 'If most new ideas are going to be attacked, the safest course is to continue doing whatever you were doing before' (Macfarquhar and Fairbank 1978b: 504). These fears lasted a long time – until the early 1980s.

The tremendous loss of human resources made it impossible to address and redress the large-scale deforestation caused by the Great Leap Forward. Without an effective Ministry of Forestry, it also became more difficult to control forest fires. From 1966 to 1977, there were 110,000 forest fires in China, resulting in a yearly loss of more than 670,000 ha of forests and a total loss of 810,000,000 $m^3$ of trees over 11 years (Wang 2000).

During the Cultural Revolution, the production of timber slowed down, but this did not stop the destruction of forests. In many cases, deforestation accelerated because the dismissal of most state employees reduced control over who could cut forests, which became "free for all". While the State Planning Commission and the MOF had previously allocated timber to different departments, deciding what

problems as the "60 articles on agriculture" (issued in March 1961 to address the mistakes made during the Great Leap Forward), but focused on forestry.

[7] For example, Heilongjiang province had about 1,600 forest technicians conducting forest surveys for the State *linchang* before the Cultural Revolution of 1966, after which there were only 177 technicians left (Yu 1989).

quantity of timber should be supplied by whom, the power of these two organizations was severely curtailed during the Cultural Revolution, and any government department could demand the trees it desired from any forest enterprise or state *linchang*. Similarly, every province in the country had the right to obtain any forest resource from any other province, sometimes without compensation. In Yunnan province, for instance, substantial quantities of timber were demanded by Hubei province with no compensation paid to Yunnan province. To prevent the loss of forest produce, many provinces chose to fell huge tracts of forest before it became allocated to other provinces.

Deforestation during the Cultural Revolution was not only the result of additional logging, lack of reforestation, and uncontrolled forest fires. As in previous decades, China's population was expanding, and it was necessary to produce increasing amounts of food. Indeed, the performance of cadres was evaluated according to the amount of grain produced. However, since farmland could only be claimed from forestland, additional production of grain led to additional deforestation. For example, between 1968 and 1978, 25–67 % of the forestland in the Counties of Baoqing, Luobei, and Suibin in northern China were transformed into farmland (Li 1988b). Moreover, most of the labor was involved in agricultural production, leaving the forests without care.

Official data indicate that forest cover dropped from 12.7 to 12 % between 1966 and 1977, although many believe the official data to be inaccurate (CFY 1987). The Cultural Revolution officially ended with the death of Mao Zedong in 1976. However, beginning from the early 1970s, its more extreme facets were slowly relaxed as former cadres (including the future leader, Deng Xiaoping) were readmitted to the government. From 1971 to 1973, many former members of the MOF returned to their old positions and began to reinstate the forest policies that they had introduced many years earlier. In this context, the MOF held a National Forestry Conference in September 1971 with the intention of addressing deforestation through extremely ambitious reforestation policies. During the 10 years from 1967 to 1978, farmers planted 830,000 ha of trees every year on average, the bulk of which was planted after 1971. The total area reforested from 1958 to 1967 was less than the area reforested every single year from 1967 to 1978, though how many of these trees survived remains unknown (Li 1985; SFA 1999).

## The Post-1978 Reforms

In 1978, China was one of the poorest countries in the world. The Great Leap Forward, the Cultural Revolution and the commune system left 250 million people under the poverty line (NBS 2004: 176). For those residing in rural areas – a classification that could be applied to 80 % of the population at the time – living conditions were particularly difficult. In 1978, the standard of living for residents of rural regions was considered below that of 1949 (Hsu 2002).

When Deng Xiaoping became China's de-facto leader in 1977, he ushered in sweeping political and economic reform, beginning in the rural areas. The challenge was that Chinese reforms could not be based on a tested blueprint, because no other country had ever experienced the conditions found in China. As a result, Deng Xiaoping's approach was to experiment with new policies in one particular province, and gradually implement them in the rest of the country if they were successful, an approach called "*crossing the river* by feeling for *stones*" (Yuan et al. 1996; Chen 1999).

From 1980 to 1984, the commune system was reorganized to "three-level administrative units". The erstwhile commune was renamed "township", its production brigade became an "administrative village" and the production team was called a "natural village."[8] While the boundaries of these political and economic entities remained more or less unchanged, the changes in name were accompanied by profound changes in their organisation and operation, especially in relation to forestry.

With regard to the forestry sector, Deng Xiaoping's government inherited a very unsustainable mode of operation. The Chinese state had already set up state-owned forest enterprises and *linchangs* to provide timber for the national economy. However, until 1979, the government had sold this timber at low prices in order to subsidize the industrial development of the country. As a consequence, insufficient funds were available for reforestation, and after 30 years of over harvesting, national forest cover had dropped significantly. Only in the late 1970s did China begin to realize that it faced a supply and demand crisis due to insufficient reforestation (Richardson 1990: 110).

Meanwhile, the government set about to reform the legal system, including legislation pertaining to forestry. Up until 1978, there was no Forest Law – only resolutions or instructions from the state addressing individual issues such as pest prevention, fire control, or reforestation. Without a unified policy, local governments adapted state guidelines to their own requirements. After 1978, the government enacted several national laws which the people could consult and which outlined the permitted and forbidden activities. They also included tougher punishments for lawbreakers.

## *The Forest Reforms of the 1980s*

The Forest Law, which began as a trial version in 1979, was officially promulgated in 1984 (Zhang 1989), making for the first time forestry laws uniform across the nation The Forest Law was revised in 1998, and this version remains in force today.

The first change brought about by the constitutional reforms of 1980–1984 was land reform, which distributed the forestland belonging to the communes to counties,

[8] Each commune had a few production brigades, and each production brigade comprised a few production teams.

villages or households. The second change directly followed and pertained to the distribution of wood. During the commune period, whenever a commune wanted wood (e.g. to build a school), it simply ordered it from the production brigade, which was responsible for felling trees to supply it (Liu 2010). With these reforms, the household, the village, and the "*xiangcun linchang*" (village *linchang*) were recognized as separate units by the county authorities; these units paid taxes to the county, and could not be told how to operate. Thus, if a county required timber to build a school, it had to buy it from a village *linchangs*, village or household, for the market price. The third change was related to the obligation imposed on counties to reforest large tracks of barren hills, and in particular the role households were now expected and encouraged to play in the management of the nation's forests. The following pages will discuss these changes in households' rights and obligations.

In China, there are two broad categories of forestland tenureship. On the one hand, in the more populated south-east, where the GfG was implemented from 1998, the forests were owned by collective farms, villages, and households. From 1984 to 1988, the first period such data is available, together they controlled 54.7 % of forestland and 19.28 % of the forest stand volume (FRSOC 1989). On the other hand, in the more sparsely inhabited north-east, state enterprises (under the management of forest enterprises or state *linchangs*) either owned or directly managed very large tracks of forests, which from 1984 to 1988 consisted of 45.3 % of forestland, and 80.72 % of the total forest stand volume. This percentage was destined to drop during the period, albeit marginally, owing to a reallocation of forest land from the state enterprises to the villages. In 1998, 41.58 % of forestland and 70.64 % of the forest stand volume were under the management of state enterprises (CFY 1999).

## *Forestland Tenure Reform in Rural Villages*

Some of the most substantial changes to China's policy related to forest management occurred after Mao Zedong's death, which led to the transformation of the commune system into the Household Contract Responsibility System (HRS). Announced in December 1978, the HRS had reached 94.5 % of Chinese households by December 1983, marking a major shift in the administrative organization of the entire country (Xiao 2008) and making a significant impact on how forests were owned and managed.

With the HRS, farmers had to supply a certain quota of specified products. Once their quota was produced, they were free to decide what else they wanted to produce, and to sell that surplus in the open market. In the beginning, farmers were given control of a piece of land for 4 months – one agricultural season – in order to test the efficacy of the policy. Deeming the experiment to be successful, the government expanded the program for 1 year, then for another 5 years and a further 15 years. Finally, in 1984, the HRS was extended for between 30 and 50 years in the barren, hilly regions (Du 2002). Since trees required time to grow, farmers were willing to plant trees on that land once they were accorded rights over the land for

15 years. To ensure that farmers benefited from the HRS, the government liberalized the sale of agricultural products from 1985 onwards. The results of the HRS were impressive, and resulted in a 61 % increase in the Chinese agricultural sector's output between 1978 and 1984 (McMillan et al. 1989).

The success of the HRS encouraged the State Council of the People's Republic of China (PRC) to issue a *Resolution on Issues Concerning Forest Protection and Development* (Feng 2005) in March 1981. The Resolution was similar in some ways to the HRS, but had a focus on forestry.[9] Known also as the "Three Fixes", the aim was to address three key issues:

1. Forestland ownership: Providing clarification on the rights to forests, with a special focus on forests in mountainous areas.
2. Mountain use rights: Delimiting the boundaries of private plots.
3. Responsibility for forest management: Establishing a forestry production responsibility system, aiming at reforesting deforested hills belonging to the farmers, and preventing fires.

The Three Fixes policy meant a return to the autonomy of individual farmers over their forests and the abandonment of the communal farming system. This was accomplished by transferring responsibility for forestland previously managed by commune *linchangs* to individual households. Land certificates were issued by the county government to households which specified their land ownership and transferred the resultant responsibility (and the benefits) of forest plantation and management to them. Not only could they now decide which tree species to plant, when to plant them, and when to harvest them, farmers could also prevent others from taking their trees and were no longer required to follow the demands of rural leaders. In this way, millions of peasants were given the freedom to plan and organize most of their agricultural and forestry production, while also enjoying the profits.

With rights came responsibilities. Land ownership also required that the farmers meet certain obligations, such as preventing forest fires, with fines levied for failing to do so. Additionally, owners could not transform forestland into farmland or cut trees "unreasonably" (CFY 1987). After the reform, the communes came to have fewer functions in rural affairs, including forest management, which was taken up directly by farmers. While local governments still took care of general administration, they no longer controlled specific actions such as when the farmers would plant and harvest crops or trees, although sales had to go through the county offices until 1984. Thus, during this period, the role of the government changed from that of direct management to that of indirectly influencing farmers' activities by supervising them, issuing guidelines, allocating quotas and determining the contract price (Liu 2006).

While there is a lack of official data for how much land was transferred to private management from collective control under the Three Fixes policy, there is no doubt that it was a substantial amount. For example, in 1975 before the reform,

[9]The policy had been slowly implemented in the mid-1970s in some areas, and it was instituted nationwide through an official national policy in 1981.

communes in Yunnan province owned 99 % of the land. Shortly following the introduction of the Three Fixes in 1982, 6 % of the land had been re-distributed; a number that grew to 55 % by 1983. However, the local governments often imposed a "use it or lose it" approach on-farmers, claiming back the land if households were unwilling or unable to plants trees on their barren land (Liu et al. 1998; Rozelle et al. 2000). As such, between 1983 and 1988, the total percentage of land distributed among village households dropped slightly. In 1995, 45 % of the land was collective, 44 % comprised responsibility forests, and 11 % consisted of private forests.

Many observers believe that, in its early phases, the Three Fixes policy did not have as positive an impact on China's forestry as the HRS had for agriculture, a situation which can be attributed to various factors (He and Zhu 2010; Liu 2007). First, by attempting to distribute forestland of different quality equally among villagers, local authorities split up the land into small parcels. Thus, many households owned two or three small land parcels located at a considerable distance both from the village and from one another. Since small, fragmented forest plots are difficult and economically inefficient to manage, few households were willing to invest in, or take care of, their plots. The sale and purchase of land was not allowed until 1998, which prevented the consolidation of forestland. There was also no symmetry or direct relationship between rights, obligations and responsibilities within the policy, and households found they had to fulfil more obligations even though their rights and benefits were limited and the economic returns were low. Also, when attempting to implement the Three Fixes policy as rapidly as possible, some local governments distributed forestland without fixing clear boundaries, leading to disputes over forest rights (He and Zhu 2010; Liu 2007).

The consequence was a chronically underperforming village forestry sector, and in some areas, the distribution of land led to widespread deforestation in the mid-1980s (Liu 2006). Scholars have proposed manifold reasons for this. Some scholars also claim that insecurity regarding the state's long-term commitment to maintaining the private tenure of forestland discouraged many farmers from investing in the productivity of their woodlots, in forest plantations, or in attempting natural forest management (Ma 1991).

The deforestation led the government to mistrust the households' ability to manage forests sustainably. In some regions, the disappointment with the Three Fixes was so great that forests that had been distributed were returned to collective management. In Jiangxi province, for instance, in 1986 over 92 % of the collective forestland was under private management, while in 2000, it had dropped to 60 % (Liu and Yuan 2007).

In spite of the problems with the Three Fixes policy, its implementation led to improved household property rights. Thus, it can be said that while in the short term the Three Fixes led to deforestation, in the long term, the reforms that the Three Fixes started were successful in reducing the speed of transformation of forestland into agricultural lands (Rozelle et al. 2000), and in increasing forest cover rates (Yin 2003; Rozelle et al. 2000).

## *The Forest Law of 1984*

In the face of large-scale deforestation, and in order to address the lack of forest investments, in 1984 a new forest law was enacted that took effect 1 January, 1985. This new forest law formalised not only the ownership of trees by households, but also household collaboration, outlining the conditions (and rights) under which households are permitted to plant trees on forestland owned by others. Moreover, under the law villagers would be permitted to pool their individual, uneconomical forests as well as invest in government or other peoples' land for the first time since 1949.

The Forest Law of 1984 also established a system of state-determined timber harvest quotas to halt deforestation. Under this system, a household has to apply to the local government for a quota to be allowed to cut trees on its land, known as the annual allowable cut. Hence, the new regulations introduced in this forest law limited the usage rights given to households a few years earlier. The difficulty farmers experienced in obtaining timber harvest quotas, without which they were not allowed to cut and sell timber, strongly reduced their potential profits. Thus, farmers' skepticism (mentioned above) as to how long the devolution of forestland tenure – and freedom to trade forest products – would really be effective (which dissuaded them from investing in reforestation) was substantiated by the Forest Law of 1984. The quota system is still in force today, and it is considered one of the main factors that reduced the degree of autonomy available to farmers as regards the sale of timber.

**Conclusions**

In this chapter, we looked at the Chinese government's policies towards the forestry sector during the period from 1949 to 1998, a period during which the country's development fuelled the vast exploitation of the forests. Regarded largely as unused farmland in China, forests were given little importance. Though compared to its predecessors, the communist government used more progressive forest management practices, including the establishment of the first Chinese Ministry of Forestry. On the other hand, the primary aim of the MOF was still to maximize timber output and increase agricultural land, at least until the early 1980s. Some efforts were made by the MOF to establish reforestation initiatives and sustainable forest use practices, but these endeavours were largely ignored by the population and the local authorities. In the face of national interest in increasing agricultural and industrial output, sustainable forestry could not compete.

The scarcity of land resources, particularly in mountainous regions of China, prompted additional massive deforestation to provide more cultivated land. Coupled with the unfavourable mountainous and hilly landscapes and uneven rainfalls, deforestation and farming on inappropriate land, in particular slope land, caused an increase in the scope and intensity of water runoff and

(continued)

soil erosion and a decline in the ecosystem's capacity of regulating water and holding soil. Excessive commercial logging and the cutting down of the forest on hillsides for cultivation in the upper and middle reaches of the basins have led to severe consequences in downstream areas. The soil- and water-eroded area has reached 360 million ha, accounting for 38.2 % of the country's land area and resulting in a soil loss of 5 billion tons annually (Lei and Zhu 2002). Increased soil erosion has silted streams, reduced hydraulic capacity of the rivers and caused higher frequencies of flooding (Smil 1993; World Wildlife Fund and State Forest Administration 2003). Records show the annual soil loss in the two rivers to be as high as 4 billion tons (World Wildlife Fund and State Forest Administration 2003).

Market economy principles were gradually introduced by the government for all agricultural products after 1978. Until 1978, the state had emphasized the production of food grains and imposed increasing production quotas on-farmers. After 1978, however, farmers were given more freedom as to which crops to grow (Xiao 2008). This promoted the cultivation of nuts, fruits, and many other economic trees, especially in hilly areas, where forestry began to be recognized as the key sector for economic development (Du 2002). Furthermore, farmers were free to sell their products in the market. In 1985, the government carried out further political and economic reforms, encouraging local counties to specialise in the production of selected products and trade those products with other counties. This was a radical reform for Communist China. With regard to forestry, this specialisation involved producing forest products for sale outside the county.

The oversupply of on-farm labor and the inaccessibility to off-farm labor market opportunities have been pointed out as major driving factors for both rural poverty and the cultivation of marginal, low yield, and highly sloped lands. Farmers in poor regions have used their greater production freedom to aggressively seek new cropping and grazing lands, often resulting in more ecologically sensitive patches on steeper slopes being claimed and degraded. Farming on steep slopes became common due to the combined effect of demographic expansion and poor regulation. This is a problem that the GfG was set to address, with its emphasis on reforesting – or returning to the natural vegetation – primarily slope land. Yields on these lands have generally remained low, which means that poverty in these regions has persisted (Du 2001). The GfG, by freeing labor from farming less productive land, with compensation for their losses, and encouraging alternative employment, also addressed the problem of poverty.

The reforms undertaken since 1978 (such as allowing the "ownership" and "sale" of forestland, trees, and the products of trees, as well as the pooling of investment and the quota system for felling trees) set the groundwork and the legal framework that made it possible to implement and operate the GfG.

(continued)

It was useful that these reforms were undertaken 15–20 years before the start of the GfG, as the farmers were able to become accustomed to the new legal framework. This is important in the Chinese context because of a history of frequent, short-lived reforms, which result in the general mistrust of government policies. Thus, the GfG (and the other reforestation and ecological restoration programs undertaken from the late 1990s/early 2000s) was a direct result of the forest policies of the previous decades. While they had to be implemented *because* of the deforestation that had occurred, they could not have been implemented in the same way *without* the reforms undertaken . The following chapter introduces the GfG by comparing it to the other reforestation and ecological restoration programs undertaken in China.

# Chapter 2
# China's Reforestation and Rural Development Programs

**Abstract** This chapter describes the six largest programs introduced during the late 1990s and early 2000s. Together, these six Key Forestry Programs (KFPs) cover 97 % of China's counties and target over 100 million ha of land for forestation. The Grain for Green is the largest of these programs, in terms of area covered, people affected, and money invested. This book only reviews the Grain for Green, but in many villages more than one program was introduced concurrently. In this chapter, we argue that the government introduced the KFPs not only because environmental deterioration had reached a critical point, but also because China was producing a surplus of grain, which lowered farmers' incomes, and because inequality between the eastern and western provinces was reaching a critical point. The Grain for Green in particular addressed all these problems concurrently, through direct payments to poor farmers willing to set aside marginal land. Partly for this reason, the Grain for Green is considered by many as the best reforestation and rural development program ever undertaken in China.

**Keywords** Reforestation policies • Rural development policies • Key Forestry Programs • Slope Land Conversion Program • Natural Forest Protection Program • Shelterbelt Development Programs • Desertification Control Program • Wildlife Conservation and Nature Reserve Development Program • Fast-Growing and High-Yielding Timber Plantation Program

## Introduction

Forest resources in China are characterised by inadequate supply, low quality and uneven geographic distribution. The period with the greatest deforestation occurred in the second half of the twentieth century, when forest coverage decreased from 30 to 40 % in 1949 to about 10 % in the late 1990s. The rapid increase in the nation's population, coupled with rapid economic development, resulted in enormous consumption of forest resources. At the turn of the millennia, the forest area of China was only taking up 4.1 % of land mass, and the stocking volume was merely 2.9 %, of the world's total (Lei 2002). This was not sufficient to meet the production and livelihood needs of a country accounting for 22 % of the world's population.



C.O. Delang, Z. Yuan, *China's Grain for Green Program*,
DOI 10.1007/978-3-319-11505-4_2

Since 1950, a total of 10 billion cubic metres of forest resources have been consumed nationwide. By 2050, the demand for forest resources is estimated to reach at least 18.5 billion cubic metres – 60 % more than the gross forest resources consumed in 2002 (370 million cubic metres).

An unfavourable physical landscape characterised by mountains and hills, uneven rainfalls, excessive commercial logging, and the cutting down of the forest on hillsides for cultivation in the upper and middle reaches of the Yangtze and Yellow river basins "caused an increase in the scope and intensity of water runoff and soil erosion and a decline in the ecosystem's capacity of regulating water and holding soil" (Yin et al. 2005: 19). The soil- and water-eroded areas reached 360 million ha, which accounted for 38.2 % of the country's total land area (more than three times the world average) and resulted in soil loss of 5 billion tonnes annually (Lei and Zhu 2002; Huang 2000). Increased soil erosion silted streams, reduced hydraulic capacity of the rivers and caused higher frequencies of flooding (Smil 1993; World Wildlife Fund and State Forest Administration 2003). Records show the annual soil loss in the two rivers to be as high as 4 billion tonnes (World Wildlife Fund and State Forest Administration 2003).

Using data from China's second soil census, Yang (1994) found that around 8 % of the country's cultivated land was affected by "intensive" water erosion, and another 26 % was affected by "light to medium" erosion. South-west China (containing the upper watershed of the Yangtze River) was estimated to contain 25 % of China's eroded land and 39 % of China's cultivated areas affected by "intensive" water erosion (Bennett 2008). According to official estimates, the soil erosion area in the Yangtze and Yellow river basins reached 75 million ha, with sediments of over 2 billion tonnes (Li 2001). On the other hand, the Loess Plateau (containing the upper watershed of the Yellow River) was estimated to contain 22 % of China's eroded land, and 19 % of China's cultivated area affected by "intensive" water erosion. Uncontrolled grazing and poor maintenance of rangelands caused extensive loss of grass cover, and contributed to soil erosion problems in the Loess Plateau. As a result of these dire conditions of forests and land, from the late 1990s the Chinese government undertook a number of forestry reforms. This chapter discusses these initiatives.

## Drivers Behind the Chinese Government's Response to Deforestation

The United Nations Conference on Environment and Development (UNCED), held in Rio de Janeiro in 1992, launched "Agenda 21", a wide-ranging, non-binding and voluntary initiative that championed the importance of sustainable resource management. Section II of the document in particular focused on reducing the extent of deforestation and protecting biodiversity. In response, the Chinese government published its own "China's Agenda 21" in 1994, which for the first time recognised the importance of protecting forests alongside the need to produce timber for the national economy, and acknowledged the need to use market approaches to address environmental problems (Lei and Zhu 2002; ACCA21 1994: 135). In the same vein, in May 1995, the forestry department published a blueprint paper, "The Forestry Executive Plan of the Agenda 21 in China" (DOF 1995). According to Lei (2002), the government also developed a "Blueprint for Ecosystem Development in China", with the objective of raising the forest cover to over 26 % by 2050, which required a net increase of 90.66 million ha in forest area. It would have taken 140 years to achieve that objective at the speed of the nationwide tree-planting campaign initiated in 1981 (Lei 2002).

In spite of these declarations and plans, little was done until the drought of the Yellow River in 1997, and the devastating floods in the middle reaches of the Yangtze River during the summer of 1998. During the drought in 1997, the Yellow River had no water discharged to the sea for 330 days, putting industrial, agricultural, and residential water uses in the northern plains in great jeopardy (Xu and Cao 2002). On the other hand, in 1998, massive floods along the Yangtze River and its tributaries claimed the lives of some 4,000 people, displaced some 18 million people and led to more than $12 billion in property damage and output losses (Lu et al. 2002). The Yellow River flows through nine provinces, and in 2000, some 110 million people lived in its basin, while another 55 million lived outside the river basin, but in areas irrigated by the Yellow River. The Yangtze River, the largest river in China, is even more important. The main channels of the Yangtze River flows through 11 provinces, and its basin accounts for some 18.75 % of China's land area, which produces 42 % of China's GDP. Both rivers basins are extremely important areas for the economic, cultural and social development of China.

Many environmental experts blamed these floods on soil erosion and deforestation upstream (World Bank 2001). There are two factors that contribute to the environmental stability of the river basins. The first one is a rich vegetation cover. Intensely forested areas with good ground vegetation cover in the upper reaches reduce the direct impact of water moving on the ground. After decades of deforestation, by the late 1990s few trees were found in the upper reaches of the rivers, even in uninhabited regions. The deforestation processes in western China were held responsible "for the increasing magnitude and frequency of floods that destroyed large areas in the middle and lower reaches of China's major rivers, such as the Yangtze and Yellow River" (Zhou 2001b). The second factor was the large lake areas. Lakes with adequate storage capacity were able to temporarily store overland flows in the watersheds. However, in recent decades, many of these flood plains and embankments around lakes were built upon, dammed, or used for agriculture, and gradually disappeared.

Following these extensive droughts and floods, the Chinese government finally instigated programs of reforestation and ecological restoration through six Key Forestry Programs (KFP). Since the extent of the disasters was in part attributed to clear cutting for agriculture on mountainous slopes, the most important programs (in terms of people involved and money invested) of the Chinese government was a nationwide reforestation program among farmers, the Grain for Green (GfG) scheme, also known as Sloping Land Conversion Program (SLCP). This program converts marginal croplands on steep hillsides and slopes into grassland and forests. The GfG is only one of the six programs, but the largest in terms of forest area, people involved, and expenditure. This book reviews the modes of operation, successes and failures of the program, but we find it useful to briefly introduce here the other five KFPs as well.

## The Six Reforestation Programs

The six KFPs cover more than 97 % of China's counties and target 76 million ha of land for afforestation (Wang et al. 2007a). The six KFPs are expected to speed up the process of restoration of the forest ecosystems, which are expected to be restored by 2050. The forest cover is expected to reach and be maintained at over 26 % to improve ecological conditions and restore the landscape (Lei and Zhu 2002). Together, these six KFPs will cost almost Yuan 1 trillion by the time they come to an end. The six forestry programs, at the core of the Chinese government's reforestation and ecological restoration efforts, are:

1. The Grain for Green Program (GfG)
2. The Natural Forest Protection Program
3. The Key Shelterbelt Development Programs in Regions such as the Three North and the Middle and Lower Reaches of the Yangtze River
4. The Sandification Control Program for Areas in the Vicinity of Beijing and Tianjin
5. The Wildlife Conservation and Nature Reserve Development Program
6. The Fast-Growing and High-Yielding Timber Plantation Development Program in Key Regions.

Lei and Zhu (2002) reported of the early successes of the programs, stating in 2002 that the area of plantations nationwide reached 46.66 million ha, which consisted in 26 % of the world's total plantation areas during that year, and made China rank first in the world. Overall, the forest area had risen to 159 million ha, the stocking volume to 11.27 billion cubic metres, and the forest cover increased from 8.6 % in the early 1950s to 16.55 % in 2002.[1]

### *The Grain for Green Program (GfG)*

Slope farming and overgrazing were the most important causes of soil erosion and desertification in western China. Xu et al. (2006a) estimated that of the 34.07 million ha of farmland in the Yangtze and Yellow river basins, 4.25 million ha were

[1] These official figures have to be taken with a pinch of salt. Until 1994 the canopy density had to be over 30 % for a vegetated area to be considered a forest, while from 1994 onwards a canopy density of 20 % was sufficient. Nevertheless, it is undeniable that the forest cover has considerably increased with the six KFPs.

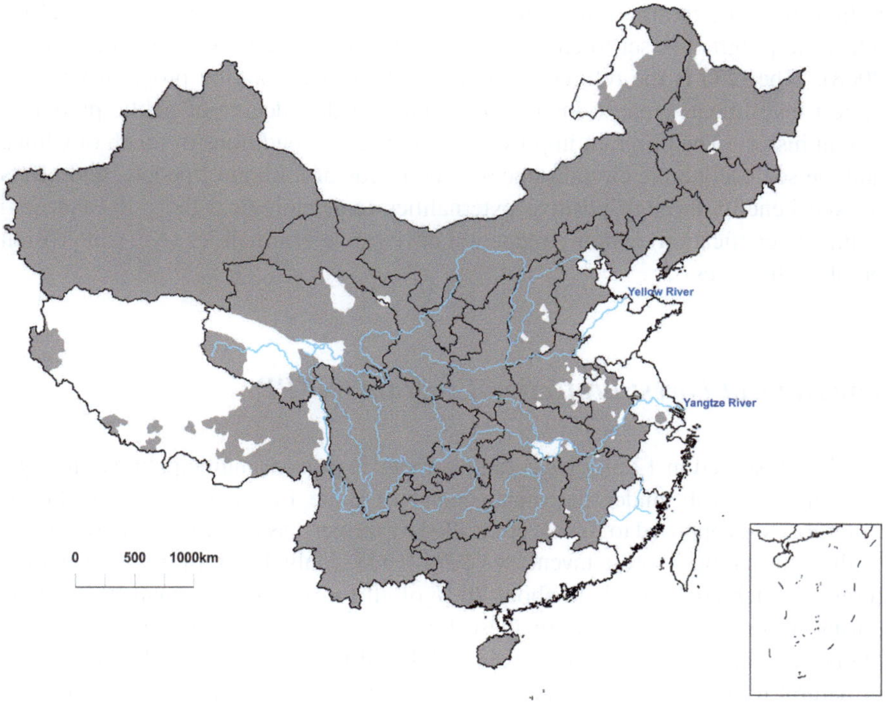


**Fig. 2.1** Areas in which the Slope Land Conversion Program was implemented (Source: Delang and Wang 2013)

on slopes of 25° or greater. It was estimated that farming on such slopes could have caused the average erosion index to reach 4,000 tonnes per km$^2$ per year. However, with proper forest coverage, 80–90 % of the erosion could have been reduced (Yin et al. 2005). The GfG was put in place primarily to reconvert steep slopes that had been cleared for farming to their original vegetation (forest or grassland), thereby reducing siltation in the rivers and alleviating farmers' poverty.

The GfG is the largest reforestation program in the world, involving 124 million people, 32 million households in a total of 1,897 counties and 25 provinces, and the Xinjiang Production and Construction Corps (Fig. 2.1) (Mao et al. 2013). Lei (2002), Deputy Chief Administrator, reported that the program planned to convert 14.66 million ha of cropland to forest, and cover 17.33 million ha of barren land with trees during the period from 2001 to 2010. Between 1999 and 2012, China actually reforested a total of 24.86 million ha through the GfG, of which 9.06 million ha was former farmland and 15.8 million ha was barren hills and wasteland suitable for forests (SFA 2013b). Upon completion of the program, the forest and grass cover of the scheme's target area would be raised by 5 %; 86.66 million ha of soil- and water-eroded area would be brought under control, and 103 million ha of sand-fixation areas would be established.

By the end of 2008, the central government had invested a cumulative total of Yuan 191.8 billion in the GfG. The plans are for further investments of Yuan 240 billion, bringing the total investment to no less than Yuan 431.8 billion by 2016, when the program is set to end (National Development and Reform Commission 2008).[2] The GfG is the reforestation and ecological restoration program with the largest investment, greatest involvement, and broadest degree of public participation in history. The program improves the ecological conditions of much of China, and the socioeconomic circumstances of hundreds of million of people, and offers off-site benefits through positive externalities (e.g. biological diversity and economic diversification) and/or a reduction of negative externalities (e.g. soil erosion or labor shortages).

## *The Natural Forest Protection Program (NFPP)*

The NFPP started in 12 provinces (autonomous regions or municipalities) in 1998, and targets forests under the management of State Linchang and State Forest Enterprises, as opposed to forests controlled by farmers, as the GfG does. According to the fifth national forest inventory (1994–1998), only 112 million ha of natural forests (which corresponds to about 70 % of all forested land) remained in China during these years, and most of these forests were degraded because of various human activities (SFA 2000c). The aims of the NFPP were to halt timber harvesting of natural forests, to protect and regenerate these forests, and to reforest, so as to meet the domestic demand for timber. To achieve the overall goal to protect and restore natural forests, the NFPP developed short, medium, and long-term goals as stepping stones (Liu et al. 2008).

The short-term goals (1998–2000) were to reduce or eliminate timber harvesting from natural forests, and to create alternative employment for those employed by forest enterprises. Commercial logging was to be completely banned in the upper reaches of the Yangtze and Yellow Rivers, as well as in Hainan Province by 2000, and substantially reduced elsewhere (Liu et al. 2008; Xu et al. 2006a). Significant steps were taken toward achieving the NFPP's short-term goals, such as generating alternative jobs for those previously employed by forest enterprises and eventually altering the employment and economic structure in forestry (Liu et al. 2008; Xu et al. 2006a). The dominant source of employment shifted from logging to forest management and plantation-farming (Liu et al. 2008).

The medium-term goals (2001–2010) were to construct and protect forests for ecological benefits and to increase the capacity for timber harvesting from plantation forests (Liu et al. 2008). Three major objectives were expected to be achieved during this period (Xi et al. 2012; Lei 2002):

1. The existing forest resources were to be protected. A logging ban was put in place on the commercial harvest of natural forests in the upper and middle reaches of the Yellow River and the upper reaches of the Yangtze River. The timber output

[2] The program is set to end 16 years after it initially started. It started at different times in different areas, but in most places it is expected to end in 2016–2019.

from natural forests in state-owned forest areas as the north-east and Inner Mongolia was reduced by 19.905 million $m^3$ (from 32 million $m^3$ in 1997 to 12 million $m^3$ by 2003), while 94.2 million ha of natural forest were brought under strict conservation (Lei 2002).

2. Afforest and reforest an additional 30.97 million ha by 2010 by means of mountain closure (prohibiting human activities, such as fuel wood collection and grazing, to allow regrowth), aerial seeding, and artificial planting, so as to facilitate sustainable logging in the future (Xiao et al. 2010; Liu et al. 2008).
3. A total of 741,000 redundant forest workers in the program area were to be redirected and relocated (Lei 2002).

The long-term goals (2011–2050) are to restore natural forests and meet domestic demand for timber in plantation forests.

Ultimately, the program was set to cover 734 counties and 167 forest industry bureaus in key state-owned forest areas in 17 provinces (autonomous regions or municipalities) in the upper reaches of the Yangtze River and the upper and middle reaches of the Yellow River, as well as the north-east and Inner Mongolia (Wu et al. 2007) (Fig. 2.2).

A total of Yuan 96.2 billion (US$11.63 billion) were assigned for NFPP-related activities from 2000 to 2010. The central government was to invest Yuan 78.4 billion (81.5 % of the total), with the remaining Yuan 17.8 billion (22.7 %) were set to come

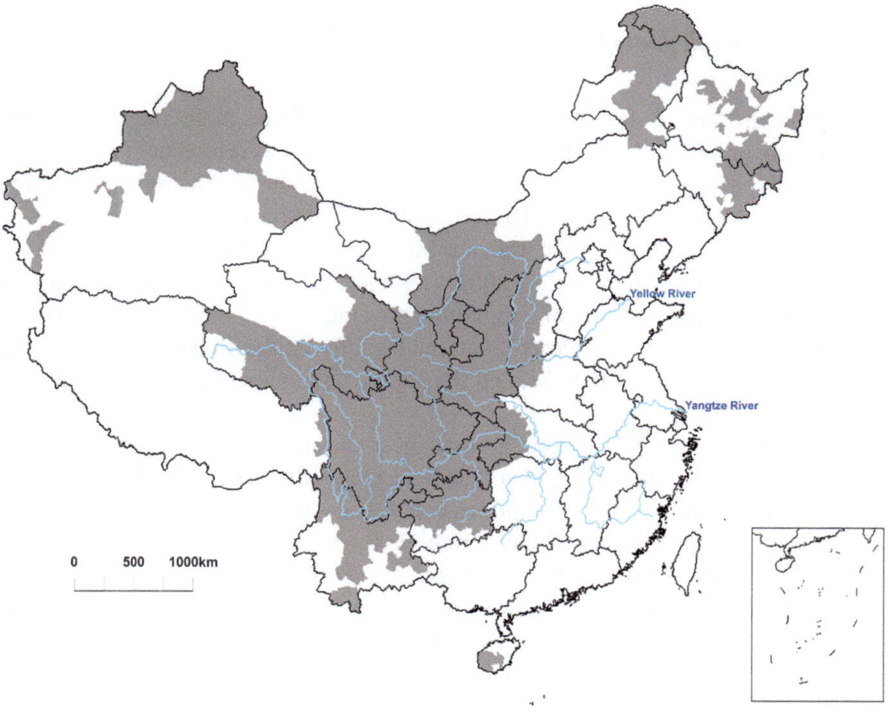


**Fig. 2.2** Areas in which the Natural Forest Protection Program was implemented (Source: Delang and Wang 2013)

from the provinces participating in the program (Yin et al. 2005). This investment was mainly used to cover the economic losses of forest enterprises that were caused by the shift from timber harvesting to tree plantations and forest management (Liu et al. 2008), including forest protection, regeneration, management, relocation of forest workers, and other related tasks (SFA 2002). In 2011 was implemented the second phase of the NFPP, which led to additional investment, a greater emphasis on forestry management, and a sharp increase (by 30.04 %) of the social security income of the remaining NFPP workers (SFA 2012b, c). NFPP is the second largest reforestation and ecological restoration program in China, in terms of the total area of implementation, and capital invested. The total amount of land converted by the GfG has exceeded the total amount of land converted by the NFPP since 2002, with the difference increasing over time (Fig. 2.3).

By 2004, 92.66 million ha of forests (60 % of the total forest area in China) were effectively managed and protected; 6.33 million ha of forests were newly established, and net stock volumes increased by 186 million $m^3$. A complete logging ban was put in place for the commercial harvest of natural forests in 13 provinces and autonomous regions along the upper reaches of the Yangtze River and the upper and middle reaches of the Yellow River. The timber output in such state-owned forest areas as the north-east and Inner Mongolia was reduced by 7.63 million $m^3$. In addition, 530,000 forest workers were redirected and resettled (Lei 2002; Xi et al. 2012).

## *The Shelterbelt Development Programs (SDP)*

The program targets desertification, mainly in the Three North region and the Yangtze River basin. The SDP is a repackaging of previous reforestation programs. Specifically, it includes the fourth phase of the "Three North" Shelterbelt Program, the second phases of the Yangtze River, Coastal and Zhuhai Shelterbelt Programs as well as the second phase of the Taihang Mountain Afforestation Program and the Plain Afforestation Program (Lei 2002; Lei and Zhu 2002). It is China's largest shelterbelt program in terms of geographic coverage, and includes "Three North" regions, coastal regions, the Pearl River, the Huai River, the Taihang Mountain, the Dongting Lake, the Poyang Lake, and the middle and lower reaches of the Yangzte River (Zhou 2001a).

The "Three North" Shelterbelt is the largest and most distinctive artificial ecological engineering project in China. The objective of the "Three North" Shelterbelt program is to "control sand and wind erosion, harness soil and water losses, improve ecological environments and produce multiple forest products" (Li et al. 2012: 71). The name "Three North" derives from the location the project is carried out. The region includes the semi-arid and arid lands in the north-east, the north and north-west of China, where desertification and erosion of soil and water constitute serious problems (Li et al. 2012; Zhu and Cheng 1994). The range of the program is enormous: 4,480 km from east to west and 560–1,460 km from south to north (Li et al. 2012). The region involves 551 counties in 13 provinces (autonomous regions or

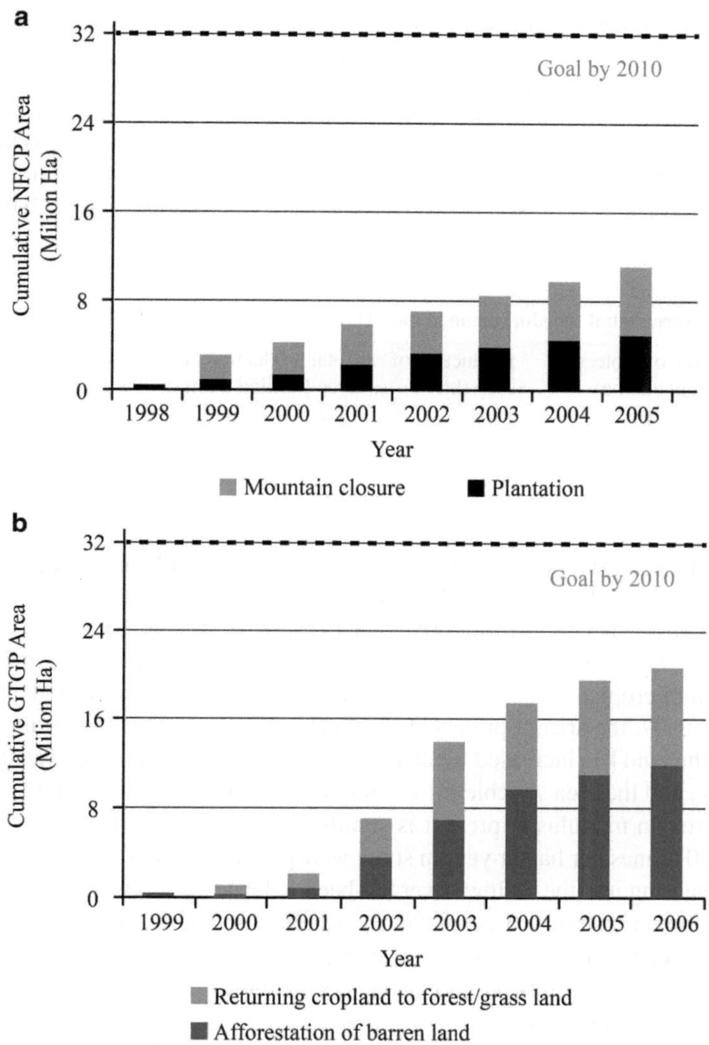


**Fig. 2.3** Cumulative amount of land under the NFPP (**a**) and GFG (**b**). The *dashed line* shows targets that had been set for 2010 (Source: Liu et al. 2008)

municipalities): Heilongjiang, Jilin, Liaoning, Hebei, Shanxi, Shaanxi, Gansu and Qinghai, Tianjin City, Beijing City, the Inner Mongolia, Ningxia and Xinjiang autonomous regions. The area is about 4.06 million km$^2$ large, and makes up 42.39 % of the total territory of the country (Li et al. 2012).

The "Three North" Shelterbelt project started in 1978 and is expected to last until 2050 (Lei 2002). The Three North scheme is meant to control desertification through a variety of measures. First, by stopping the advance of the desert. As such, about 1,060 km$^2$ of desert per year are transformed from mobile dunes to semi-fixed or fixed dunes (Zha and Gao 1997, in Li et al. 2012). Second, through afforestation.

**Table 2.1** Socio-economic transformation in the "Three North" regions in 1977 and 2007

| Year | Forest volume (million $m^3$) | Area of economic forest (million ha) | Production of fresh and dry fruits (million tonnes) | Tourist number (million persons) | Number of employment positions |
|---|---|---|---|---|---|
| 1977 | 720 | 1.8 | 7.2 | 2.3 | 110,000 |
| 2007 | 1,390 | 5.8 | 36 | 90 | 700,000 |

Source: Li et al. (2012)

**Table 2.2** Agricultural transformation in the "Three North" regions in 1977 and 2007

| Year | Area of protected cropland (ha) | Production of crop (kg/ha) | Total production of crop (million tonnes) | Area of pasture in sandy land (million ha) |
|---|---|---|---|---|
| 1977 | 505.0 | 1,770 | 60 | 4.3 |
| 2007 | 2,248.6 | 4,665 | 153 | 19.64 |

Source: Li et al. (2012)

Lei and Zhu (2002) reported that 94,600 $km^2$ of land were planned to be afforested, and 13,000 $km^2$ of desertified land brought under control from 2001 to 2010. Indeed, the forest cover increased from 5.05 % in 1978 to 10.51 % in 2008 (Table 2.1) (Li et al. 2012). By 2050, 305,800 $km^2$ are planned to be afforested.[3] Third, through the protection of cropland, and increase in crop production and pasture areas. Between 1977 and 2007, the area of protected cropland has increased by 445 %, the productivity of the land has increased by 264 %, the total agricultural output has increased by 255 %, and the area suitable for pasture has increased by 457 % (Table 2.2).

Soil erosion modulus at present is significantly lower than that in 1977, at less than 1,000 tonnes per ha per year in some well-planted areas. Similarly, the amount of sand entering into the Yellow River has been reduced by about 300 million tonnes according to one estimate (Li et al. 2012). Amelioration of the environment has also had further economic benefits. For example, it has stimulated the development of tourism, which provides employment opportunities for the local population (Table 2.1) (Li et al. 2012).

The shelterbelt development project in the middle and lower reaches of the Yangtze River involves relevant areas in 31 provinces (autonomous regions and municipalities). It was expected that 180,000 $km^2$ of land would be afforested, 73,300 $km^2$ of low-efficiency shelterbelt improved, and 373,300 $km^2$ of existing forests properly managed and protected during the period from 2001 to 2010 (Lei and Zhu 2002). The coastal shelterbelt project involves 195 counties in 11 provinces. The planned afforestation area is 35,600 $km^2$. The regional forest cover was raised from 21.7 % in 1987 to 29.1 % after 15 years of implementation (Li 2004b;

[3] When considering these changes, one should bear in mind that until 1994, the canopy density had to be over 30 % for a vegetated area to be considered a forest, while from 1994 onwards a canopy density of 20 % was sufficient.

Wenhua 2004). Finally, the Taihang Mountains afforestation project involves 110 counties in Beijing, Hebei, Henan and Shanxi Provinces. Through this project, 35,600 km$^2$ of forests are planned to be planted by 2050 (Li 2004b).

## *The Desertification Control Program (DC)*

The Desertification Control Program targets sandstorms in areas surrounding Beijing. The program covers 75 counties, with a total area of 460,000 km$^2$, in five provinces (autonomous regions or municipalities), including Beijing, Tianjin, Hebei, Shanxi and Inner Mongolia (Lei 2002). Lei and Zhu (2002) reported that during the period from 2001 to 2010, it was expected that 26,300 km$^2$ of cropland would be converted to forest, 49,400 km$^2$ of plantations would be established, 106,300 km$^2$ of grassland would be harnessed, 113,800 supporting water conservation facilities would be developed, 23,000 km$^2$ of catchment would be managed and 180,000 people would be relocated for ecological reasons. Zhou (2001a, b) reported that upon completion of the program the ecosystem in the areas in the vicinity of Beijing and Tianjin would be remarkably improved, with the forest cover expected to reach 21.4 % by 2010, an increase of 14.7 % (Lei 2002).

## *The Wildlife Conservation and Nature Reserve Development Program (WCNR)*

The Wildlife Conservation and Nature Reserve Development Program targeted such issues as species, nature and wetland protection. Priorities, between 2001 and 2010, were given to the following (Xi et al. 2012):

1. Setting up 15 wild fauna and flora protection projects (including the Giant Panda, Golden Monkey, Tibetan Antelope and plants in the orchid family) (Stanturf et al. 2012)
2. Establishing 200 nature reserve projects in the types of forest, desertified land and wetland ecosystem, 32 wetland conservation and wise use demonstration projects and 50,000 nature reserve districts (Lei 2002);
3. Completing the germplasm pools for conservation of wild fauna and flora, the national research system of wild fauna and flora and relevant monitoring networks (Lei 2002; Sun and Liqiao 2006).

Between 2001 and 2006, 831 natural reserves were created, and 19.5 million ha of forestland and special sites were protected under this program. By 2010, the number of nature reserves was set to reach 1,800, which was to include 220 nature reserves at national level, with the total area of nature reserves taking up 16.14 % of the country's land area (Lei 2002).

### *The Fast-Growing and High-Yielding Timber Plantation Program (FHTP)*

The Fast-Growing and High-Yielding Timber Plantation Program aims to resolve the supply of timber, while at the same time mitigating the pressure of timber demand on forest resources. The program covers 114 forestry bureaus (or farms) and 886 counties in 18 provinces and autonomous regions, located to the east of the isohyet of 400 mm in China (Lei 2002). Between 2001 and 2015, it plans to establish 13.33 million ha of fast-growing and high-yielding plantations. Upon completion, the program would provide 130 million $m^3$ of timber annually, accounting for 40 % of China's commercial timber consumption, thus helping to create a balance between timber supply and demand (Zhou 2001b).

## Why the Key Forestry Programs?

While the six KFPs were spurred by the drought and flooding of 1997 and 1998, it is also important to recognize that other factors may have contributed to the Chinese government starting these programs at that time. The economic development experienced in previous years had increased the costs of flooding, justifying the investment of such large amounts of money in forest conservation and reforestation. However, at the same time, it also made it possible for the Chinese government to invest such large amounts of money. Thus, while the drought and flooding spurred the urgency of investing very large amounts of money in addressing the deforestation that had taken place during the past decades, other factors also prompted the government to carry out these reforms, and allowed it to start these programs at that particular time.

First, the Chinese government wanted to become more influential in the global arena, entrenching China's status as a superpower, not only economically but also socially and environmentally, to improve its image, increase its influence, and prove that it was a good, responsible member of the world community. For example, China subscribed to the nuclear Non-Proliferation Treaty (1992) and a number of international environmental treaties, such as the Convention on Biological Diversity (1992), the Ramsar Convention (1992), the United Nations Convention to Combat Desertification in those Countries Experiencing Serious Drought and/or Desertification, Particularly in Africa (1997), the Convention on International Trade in Endangered Species of Wild Fauna and Flora (CITES 1981), the International Tropical Timber Agreement (1986), and The Convention on the Protection of New Plant Varieties (1999), all implemented by the Forestry Department (Lu 2011; Chen and Shou 2001; Zhang 1998).

Second, the need for reforms in rural western China had become more urgent. By the late 1990s, inequality between urban and rural households, in particular between the more industrialised east and the more agricultural west, had increased,

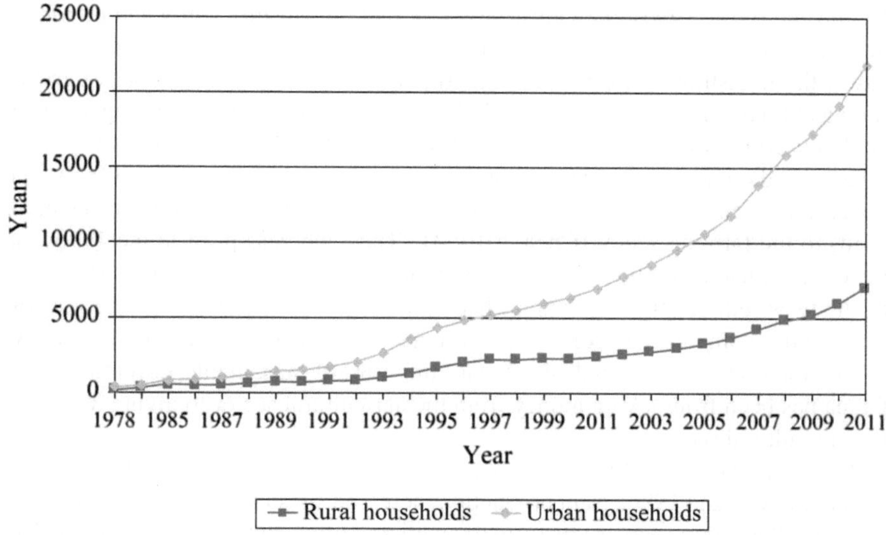


**Fig. 2.4** Annual per capita income of urban and rural household, 1978–2011 (Source: National Bureau of Statistics of China 1996–2012)

generating growing anger among the rural population that had not been able to benefit from the economic growth (Fig. 2.4). Since the majority (63.91 % in 2001, NBS 2004: 174) of the Chinese population were rural dwellers, the increasing inequality was of concern to the leaders, as those rural dwellers could generate political instability, which would imperil economic growth. In 2001, the government identified and set out to address "Three rural issues concerning agriculture, countryside, and farmers"[4]: farmers' incomes were very low while their burden was high; production was low and shrinking; rural areas lacked public services (Chen 2010; Wang 2008). These problems were to be addressed by the GfG (Zhao 2010; Zhang 2010), but also to a lesser extent by the NFPP. For example, by limiting the amount of timber logged in nationally-owned forests, the NFPP also benefited timber-producing households (Delang and Wang 2013).

Third, these reforms were made possible by the increasing wealth of the Chinese government. The Chinese economy developed enormously during previous decades, and all expectations were for the economy to continue growing (which it did, especially after China joined the WTO in 2001). The Chinese government had large revenues, which it could invest in the development of the country. Before the 1990s, the government extracted money from rural areas to help the industrial sector develop, first by taxing agricultural and forestry products and diverting the funds to the industrial sector (Wang and Delang 2011), and second by lowering the price of agricultural and forestry products, which allowed industrial workers to subsist on lower salaries and provided cheap inputs for industries (Wang and Delang 2011).

[4] 三农"问题

By the end of the 1990s, the industrial sector had grown sufficiently and no longer needed funds extracted from the agricultural and forestry sectors. Also, in some cases, the industrial sector moved upmarket and no longer needed cheap agricultural and forest products, while the salaries of the workers were increasing and they could afford to pay more for their food. Whereas in the 1950s 39 % of government revenue had come from the agricultural tax, by 2004 only 1 % did (Li 2009a). The government could afford to lower the taxation level of the farmers and allocate more funds to the farming sector (Chen 2010: 4).[5] Thus, whereas prior to the 1990s the government had taxed rural areas to help industrialise the country, towards the end of the twentieth century the country had sufficiently industrialised for the government to use some of the taxes levied in industrial areas to develop rural areas. As O'Connor (2000) concluded, the great shift toward sustainable development in China has been from taking money from the agricultural sector to giving money to the agricultural sector.

Fourth, in the late 1990s, a crisis of grain supply exceeding demand developed in China. In 1995 and 1996, grain production increased substantially, and beginning in autumn 1996, there was a fall in market prices. Abundant harvests continued in 1997 and 1998 (Tan and Chen 1999). Peasants suffered because of low prices, and China suffered because it had to provide peasants with price subsidies. Through the GfG, less productive – and prone to erosion – marginal farmland was set aside, thereby also reducing the surplus of rice produced (Delang and Wang 2013). This will be further discussed in Chap. 3.

Fifth, the considerable economic development of the country also made it possible – and more profitable – for China to shift from producing timber to importing it. Until the 1970s, China imported little timber because of political constraints (the Cultural Revolution and the Great Leap Forward) (Wang and Delang 2011). From the 1970s to 1997, it imported little timber because timber produced in China was cheaper – mainly because the price was depressed by the government (Delang and Wang 2012). In the late 1990s, it became obvious that logging generated externalities which sharply increased the cost of timber, and it made economic sense to protect the national forests and import timber from other countries. China shifted from logging its own forests to import raw wood logged abroad (Fig. 2.5).

## New Forestry Paradigms in China

The execution of these Key Forestry Programs has been a landmark in China, helping the sector to enter into a new era of renewal. Over this period, emphasis has been given to ecological, social, and economical benefits. This development is expected to propel five historic transformations.

[5] In 2000, the government started the financial reforms for agricultural products (which covered the whole national territory only in 2006). Through this reform, taxes on agricultural products, including trees and timber, were abolished. Since 2007, farmers in China no longer have to pay taxes or fees for agricultural or non-timber forest products, though they still pay limited fees for timber.

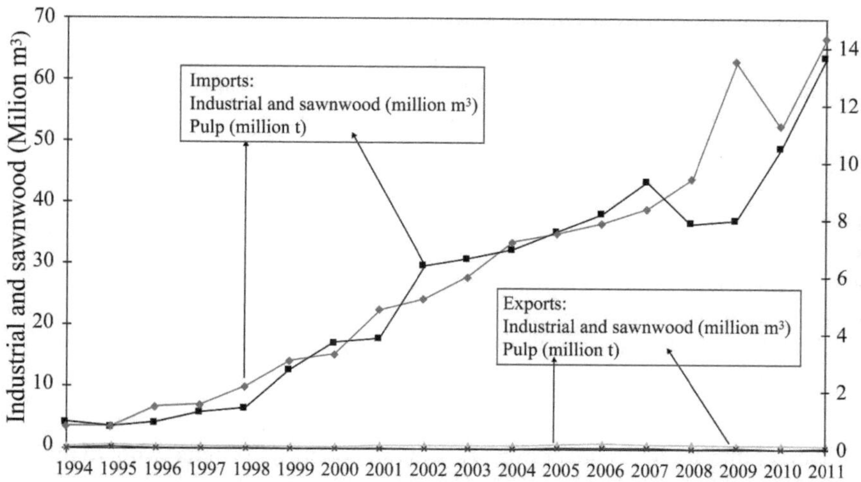


**Fig. 2.5** Imports and exports of industrial and sawn wood (1994–2011) (Source: Delang and Wang 2013)

First, the transition from considering forestry a sector of the economy that contributes to the nation's GDP, to paying more attention to the environmental benefits it provides, and its role in the sustainable development of the country.

Second, a shift from ignoring the contribution that the ecological functions of forests make to the national economy, to fully accounting for these functions. In 2001, the MOF and the Ministry of Finance agreed to establish pilot projects in 660 counties within 11 provinces and 24 nature reserves, which represented a total area of more than 13.33 million ha. The implementation of this pilot scheme helped setting in motion a new stage in which the economic value of the ecological functions of forests would be properly incorporate into economic planning.

Third, a transition from transforming forests into farmland to transforming farmland into forests.

Fourth, shifting from the previous emphasis on felling natural forests to gradually harvesting plantations. In the past, natural forests were the most important timber production areas in China. With the implementation of six KFPs, natural forests have become strictly protected. At the same time, the timber output of natural forests has been greatly reduced and the proportion of timber output of plantations been increased (Lei 2002). However, for the time being, the drop in logged forests is compensated with increased imports, which means that China has replaced logging its own forests with logging other countries' forests. Figure 2.5 shows how the imports of industrial timber and pulp have expanded more than 15 fold, while exports have remained marginal. Nevertheless, timber production will gradually shift from natural forests to plantations, and eventually all timber is expected to come from plantations.

Sixth, moving from land managed by the forestry industry for the production of timber, to forestry managed by different sectors, with input by several industries, for the benefit of the whole of China.

**Table 2.3** Comparison of six key forestry programs

| Program | Planned investment (billion Yuan) | Forestation area (10,000 ha) | Provinces covered (1,000 km$^2$) | Duration (years) |
|---|---|---|---|---|
| GfG | 361.8 | 3,328 | 9,560 | 20 |
| WCNR | 135.7 | 4,980 | 9,570 | 50 |
| NFPP | 101.8 | 10,768 | 7,700 | 13 |
| FHTP | 71.8 | 1,333 | 4,210 | 17 |
| SDP | 94.6 | 6,870 | 9,570 | 73 |
| DC | 58.4 | 2,104 | 1,530 | 11 |

Sources: Li and Zhai (2002)
Note: The data reflect the actual completion for completed projects and planned target for uncompleted projects.

Together, the six KFPs cover over 97 % of all counties in China, with 76 million ha of plantations planned to be established. This makes the programs unprecedented in history due to their wide range, large scale and great investments (Table 2.3).

Between 1998 and 2003, five of the six programs were financed mainly by the state, which covered 83.5 % of the total investment. The only exception was the market-oriented FHTP, which was financed primarily by farmers, forest enterprises and foreign capital. The state covered only 6.6 % of the total investment in that program over this 5-year period. On the other hand, the NFPP is financed by the state, and directly implemented by either state-owned forest enterprises or local forest authorities, in a rather top-down fashion. Finally, while the GfG is financed predominantly by the state, it uses a public payment scheme that directly engages millions of rural households as core agents of project implementation (Bennett 2008). Thus, in terms of decentralization and grassroots participation, the GfG is a novel program, representing an important departure from the way China has been managing its forest resources (Bennett 2008). We can also conclude that although the six KFPs have been introduced during the same period, they adopted different organisation and financial arrangement, suited to the conditions and recipients among which they were implemented.

Liu et al. (2010) address the impact of six KFPs on-farmers' income and poverty status, using a fixed-effects model and a panel dataset of 1,968 households across four provinces for ten consecutive years, between 1995 and 2004. The findings suggested that the impact of the six KFPs were mixed. The GfG, the KSD, and the NFPP had a positive impact, with the GfG having by far the greatest impact. However, the WCNR and the DC had not yet had a pronounced overall effect owing to the short time span they had been implemented, even though they may have exerted certain influence at the margin. Notably, the impact of the WCNR, if any, was negative. On the other hand, from an environmental point of view, the implementation of the NFPP has effectively protected the state natural forest resource, while the large-scale GfG has become the major driving force of the recent growth in forest resources.

The scale of GfG makes the program one of the world's largest conservation projects. Statistics of the MOF suggest that forest cover within the GfG region increased by 2 % over 8 years (Liu et al. 2008). However, the GfG is much more than a forest conservation project. It also has important socio-economic implications, and it has generated more positive socio-economic impacts than the NFPP, or any of the other forestry programs. Unlike the NFPP for example, which has cut off income from timber harvesting for many forest workers, the GfG has helped alleviate poverty through the direct subsidies it gives to farmers willing to set aside their land (see Part III). The GfG has directly benefited 120 million-farmers in more than 30 million households nationwide, whereas the NFPP has directly affected only hundreds of state-owned forest enterprises and only indirectly impacted a larger number of households (Liu et al. 2008).

The NFPP and GfG also have important global implications, although they were initially developed to address pressing environmental problems in China. As Liu and Diamond (2005) write, "China's achievements of developed-world consumption standards would approximately double the world's human resource use and environmental impact. But it is doubtful whether even the current human resource use and impact on the world could be sustained. […] China's environmental problems are therefore the world's" (Liu and Diamond 2005: 1181). Further, Liu et al. (2008) pointed out that "if implemented adequately and sustainably, these two programs could generate many benefits to China and the rest of the world by addressing a wide array of environmental issues (e.g. biodiversity loss, climate change, desertification, droughts, floods, soil erosion, and water runoff) as well as socioeconomic challenges (e.g. poverty alleviation, social conflicts, and economic development)".

**Conclusions**

The successive occurrences of ecological disasters in the late 1990s indicated that while there had been scattered, local-level successes in protecting forest ecosystems, they were overwhelmed by the worsening of the overall situation. This means that more decisive and forceful measures were needed to halt the environmental problems (Yin et al. 2005). This chapter has given an introduction to the key forest policies introduced in China in the late 1990s. Through the forestry reforms of 1998 and the six KFPs, China put in place the framework to transform the ways in which forests are managed, expoited, and protected. The aims of these programs have been lofty and well-intentioned, with some of the programs overshadowing all other reforestation and ecological restoration programs worldwide. To a large extent, the programs have been successful at reversing the deforestation, soil erosion, and desertification that had occurred during the previous decades, even though some programs have been more successful than others. In particular among farmers, the GfG is often considered to be the best reforestation program that the government has ever undertaken, largely because its objectives are not only to reforest and restore the ecological integrity of the areas, but also to alleviate poverty, and since payments are made directly to farmers who set aside their land, this latter objective is usually fulfilled.

# Part II
# Overview of the Grain for Green

This second part provides an introduction to the Grain for Green program, looking at the ways in which it was introduced and implemented, the rules that govern it, and how it operates. Most of the publications we review are from the 2000s, because less has been published over the last years. The 2000s (in particular from 2002 to 2004) were the years when much of the program was implemented, so this is only a minor drawback. However, it is indicative of the reduced interest in the GfG over the last few years, despite the fact that the GfG subsidies will come to an end starting in 2015. The end of the GfG subsidies will potentially have a considerable impact on China's rural areas, which will be discussed in Chap. 13.

Chapter 3 discusses program timeline, looking at the ways in which the program was expanded nation-wide in 2000 after being tested in three provinces – Sichuan, Shaanxi and Gansu – in 1999. In particular, the GfG was slowed down after 2003 because of fear (later found to be unfounded) over the impact that the GfG had had on food supply, and because more land had been converted over the previous 2–3 years than originally planned: often local civil servants converted more land than that allocated by the central government for a particular area, because of the very generous funds they would receive for their impoverished farmers. The chapter also gives some examples of the expansion of the program in particular counties and cities.

Chapter 4 looks at the level of farmers' compensation for all three plant types: economic trees, ecological trees, and grassland. GfG regulations stipulate that compensation should only be paid if a large number of planted trees and grasses survive (initially 70 % in the Yellow River watershed and north China, and 85 % in the Yangtze River watershed and south China, later standardized to 75 % nation-wide). We show, however, that often compensation was also given if a smaller number of trees survived. The chapter also looks at the extent to which the funds are actually delivered to the farmers, which was a concern to the farmers when the program was introduced. Finally, the chapter reviews the total incomes of the farmers, comparing pre-GfG incomes to post-GfG incomes from the same land. In more cases than not, the post-GfG incomes are higher than the pre-GfG incomes from crop cultivation, which means that the program raised farmers' incomes. However, the GfG could have converted more land with the same budget, or the same amount of land with a lower budget.

Chapter 5 reviews land selection. According to program guidelines, slope land or unproductive land prone to soil erosion should have been converted first. Evidence shows that although this has happened in many cases, it was not universal. In some cases, productive land with low slope was also converted. This might have been because program managers preferred to convert large tracks of adjacent land, and flat productive land might have been found between steeply sloped land.

Chapter 6 discusses plant selection. The GfG promotes the planting of either economic trees (trees from which a regular income may be obtained from the sale of non-timber products, such as fruits), ecological trees (trees that may be logged), or grassland. More farmers prefer to plant economic trees, because they generate higher and more regular incomes than ecological trees. However, the national standard is for ecological trees to make up 80 % of the total, and this is generally adhered to. In many places, farmers also claim that they do not have a choice of which plants to grow, but can only select from a few species.

Chapter 7 looks at household attitudes and engagement. In most places not all those who joined the program claim that they did so voluntarily, though many farmers were willing to convert their least productive land, especially when they had a surplus, and their remaining land was sufficient to grow enough food for subsistence. On the other hand, most researchers found that the GfG is now a very popular program, since the funds are rather generous and the payments regular, and there is a visible improvement in the ecological conditions of the areas where it has been implemented.

Chapter 8 discusses the institutional context within which the GfG was set up and operates, and the role of each level of government (national, provincial, prefectural, county, township) in its implementation. The implementation of the GfG is complicated by the fact that China has traditionally had a very centralized political structure, with decisions being made in Beijing and little inputs from the regional and local governments. By contrast, the GfG is a relatively decentralized program, with important decisions made at the grassroots level. The organization of such a large program, involving over 30 million households in 1,897 counties nationwide, is bound to face problems at the planning and implementation stages, and we review some of these problems in this chapter.

# Chapter 3
# Program Timeline

**Abstract** This chapter discusses program timeline, looking at the ways in which the program was expanded nation-wide in 2000 after being tested in three provinces – Sichuan, Shaanxi and Gansu – in 1999. In particular, the Grain for Green was slowed down after 2003 because of fear (later found to be unfounded) over the impact that the Grain for Green had had on food supply, and because more land had been converted over the previous 2–3 years than originally planned: often local civil servants converted more land than that allocated by the central government for a particular area, because of the very generous funds they would receive for their impoverished farmers. The chapter also gives some examples of the expansion of the program in particular counties and cities.

**Keywords** Pilot-phase • Nation-wide implementation • Program slowdown • Grain output • Price of grain

## Introduction

The GfG started in 1999 in a trial format in three western provinces: Sichuan, Gansu, and Shaanxi. Given its initial success, it was gradually expanded in the whole of China, except the eastern coastal provinces. This chapter introduces two aspects of the GfG. In the first part, we look at the geographic and temporal expansion of the program, and examine its implementation and development tracing its progress from its inception in 1999 up to the late 2000s. We also note reasons for land conversion rates to vary with time, and in particular for the slowdown in program expansion after 2004. In the second part of this chapter, we look at the situation of particular counties and cities.

## Program Timeline

The GfG began as a pilot set-aside program in late 1999 when Sichuan, Gansu, and Shaanxi provinces first conducted trial projects. During this trial period, the central government did not set any targets. Instead, the local governments had full



C.O. Delang, Z. Yuan, *China's Grain for Green Program*,
DOI 10.1007/978-3-319-11505-4_3

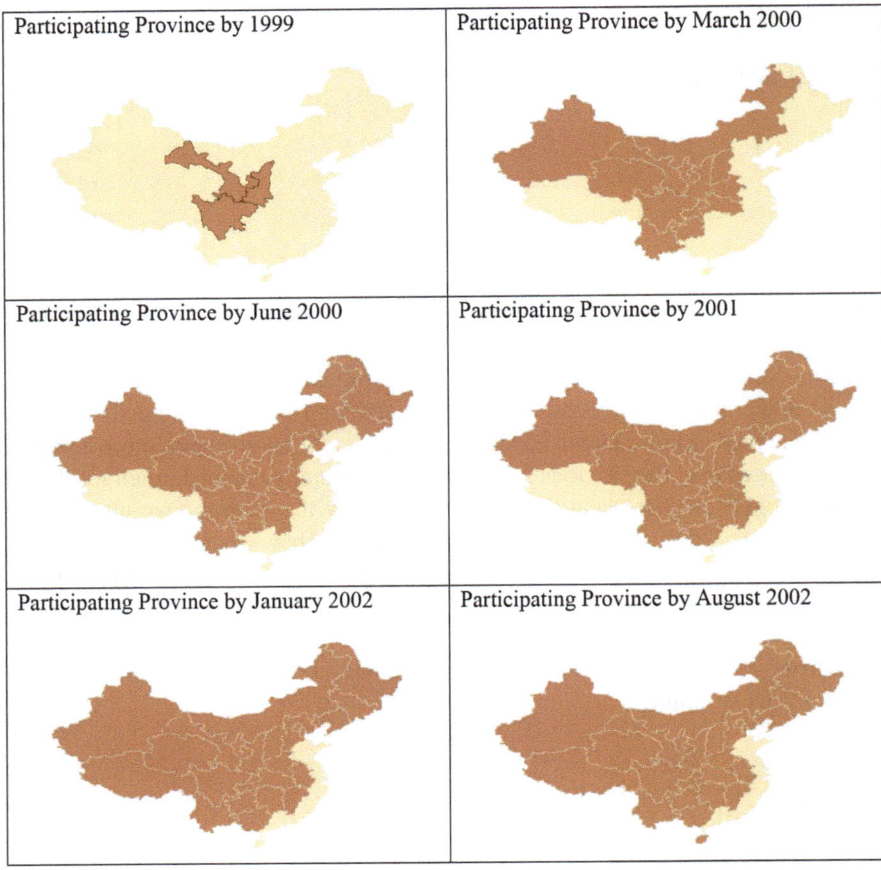


**Fig. 3.1** Participation of province by year (Source: Cui 2009)

autonomy to convert land according to their capabilities (Cui 2009). In March 2000, the GfG pilot program officially started under the ratification of the General Office of the State Council. At that time, the geographic coverage of the GfG was extended to 13 provinces: five provinces in the upper Yangtze River basin (Hubei, Chongqing, Sichuan, Guizhou, Yunnan) and eight provinces in the upper and middle Yellow River basin (Shanxi, Inner Mongolia, Henan, Shaanxi, Gansu, Qinghai, Ningxia, Xinjiang). In 2001, the GfG was further expanded to include a total of 20 provinces and autonomous regions, consisting of 400 counties and 27,000 villages. In the program villages, more than 15 million-farmers set aside their land and received payments during the pilot phase from 1999 to 2001 (Uchida et al. 2005). During the pilot phase, an average of 408,000 ha of cropland was converted per year, for a total of 1.2 million ha, and a total cost of Yuan 3.65 billion (Xu and Cao 2002). Figure 3.1 shows the provinces which were gradually included in the GfG.

The success of the pilot phase motivated the State Council to officially launch the GfG nationwide. In 2002, the program was implemented in 25 provinces and

**Table 3.1** Participating provinces of GfG implementation

| Year and month | 1999 | 2000 Mar | 2000 Jun | 2001 | 2002 Jan | 2002 Mar onwards |
|---|---|---|---|---|---|---|
| No. of participating provinces and autonomous regions | 3 | 13 | 17 | 20 | 24 | 25 |

Source: Li (2005), SFA (2003e)

**Table 3.2** GfG implementation

| Phase | | Pilot | Official implementation | Adjustment |
|---|---|---|---|---|
| Year | | 1999–2001 | 2002–2003 | 2004–2005 |
| No. of participating | Provinces and autonomous regions | 20 (1) | 25 (1) | 25 (1) |
| | Counties | 400 (1) | | 1,897 (1) |
| | Township | 5,700 (1) | 20,000 (2) | |
| | Villages | 27,000 (1) | | |
| | Households | 4.1 million (1) | | >20 million (1) |
| | Farmers | 16 million | | >97 million (1) |
| Converted cropland (10,000 ha) | | 120.61 (1) | 570.26 (3) | 187.78 (3) |
| Converted wasteland (10,000 ha) | | 109.73 (1) | 604.94 (3) | 388.33 (3) |
| Expenditure | Grain subsidy (billion kg) | 3.57 billion kg or 4.99 billion Yuan (4) | 14.89 (3) | 37.64 (3) |
| | Cash subsidy (billion Yuan) | 0.59 (4) | 3.10 (5) | 2.58 (5) |
| | Seedlings subsidy (billion Yuan) | 1.73 (4) | 8.29 (5) | 5.05 (5) |
| | Technology input (billion Yuan) | 0.026 (4) | 0.60 (5) | 0.29 (5) |
| | Total (billion Yuan) | 7.68 (4) | 31.92 (5) | 46.52 (5) |

Source: (1) Cui (2009); (2) Zhang et al. (2008a); (3) SFA (2003–2006); (4) Li (2004a); (5) Ke (2007) (exclude the cropland conversion from reforestation project DC)

autonomous regions (Table 3.1). While only five provinces were added to the program, this does not mean that the expansion was small. The number of counties enrolled in the program increased by 374 % between the end of 2001 and the end of 2002. At the end of 2002, the GfG was extended to a total of 1,897 counties, and by the end of 2003, the program had been implemented in more than 2,000 counties in 25 provinces (Table 3.2).

Under the program, the MOF planned to convert around 14.67 million ha of fragile cropland to forest by 2010. However, as the program was implemented full-scale nationwide, it expanded very rapidly. According to Bennett (2008), while during the pilot phase (1999–2001) an average of 408,000 ha of cropland was converted yearly, upon full-scale implementation beginning in 2002, the area of converted land jumped to 2.9 million ha per year in 2002 and 2003 – more six times what was achieved during the pilot phase (Bennett 2008).[1] By the end of 2003, 6.44 million ha of cropland had been converted, and 6.89 million ha of barren land had been afforested, as claimed by the MOF (SFA 2004b). In 2003 alone, a record of 6.84 million ha were afforested under the GfG (SFA 2004b). The figures provided by Bennett (2008) and the SFA (2004b) are somewhat different from those presented in Table 3.2, because they rely on different sources.

The speed of the program's expansion and land conversion once the GfG had been implemented nationwide suggests that the GfG was rather enthusiastically adopted by the local authorities and the farmers concerned. In particular, local governments were eager to receive funding from the central government, and pushed to expand the area allocated for the GfG, in many cases setting aside more land than had been allocated to that area by the central government. As a consequence, within 4 years of implementation, half of the 12-year goal had already been realised. Indeed, Bennett (2008) argues that "local governments have found ways to milk the system by focusing their efforts on increasing their land conversion quotas, either through direct negotiation, or by first overreaching their land conversion quotas and then bargaining for more subsidies". Xu et al. (2006c) also report that local governments exceeded their quotas because they benefited from distributing subsidies to the farmers. Li et al. (2006) also reported that villages with connections to local forest bureaus were able to obtain larger retirement quotas. Bennett (2008) argued that such behaviour had existed since the beginning of the program, when the three pilot provinces of Sichuan, Shaanxi and Gansu overshot their quotas by more than 100 % within 3–4 months. This continued through 2000, when 312 counties initiated land conversions on their own initiative, despite the fact that the central government's plan was to implement the pilot program in only 174 counties. Bennett (2008) states that the MOF continued to receive numerous requests from local governments asking for higher land conversion quotas, up to the late 2000s.

## *Program Slowdown*

In Chap. 2, we argued that one of the reasons for starting the GfG was that grain production had increased substantially from 1995 to 1998, resulting in a fall in market prices (Tan and Chen 1999). Peasants suffered because of low prices, while China suffered because it had to provide peasants with price subsidies. However, a shortage of grain appeared in June 2003, which was feared to be a result of the excessive conversion of farmland into forestland or grassland. This fear later proved to be unfounded (see Chap. 10), but it prompted the State Council to make a structural adjustment to the GfG – shifting the emphasis from area expansion to the maintenance of the converted land. As a result, a much lower conversion target was

[1] In 2002, 95.1 % of the cropland and 93.5 % of the devastated land targeted for afforestation or reforestation for that particular year was afforested (Bennett 2008).

**Table 3.3** Size of GfG-converted area (10,000 ha)

| Year | Total reforested Converted | Reforested cropland Converted | Reforested Barren land Converted |
|---|---|---|---|
| 1999 | 29.30 | – | – |
| 2000 | 68.36 | – | – |
| 2001 | 87.08 | 38.63 | 48.45 |
| 2002 | 442.36 | 203.98 | 238.38 |
| 2003 | 684.09 | 341.81 | 342.28 |
| 2004 | 356.82 | 101.66 | 267.49 |
| 2005 | 335.31 | 86.12 | 249.73 |
| 2006 | 120.43 | 26.89 | 93.54 |
| 2007 | 112.47 | 8.53 | 103.94 |
| 2008 | 130.73 | 1.20 | 129.54 |
| 2009 | 89.86 | 0.07 | 89.79 |
| 2010 | 99.65 | 0.03 | 99.62 |
| 2011 | 73.02 | – | – |
| Total | | 906.26 | 1,607.04 |

Source: SFA (2000–2012)

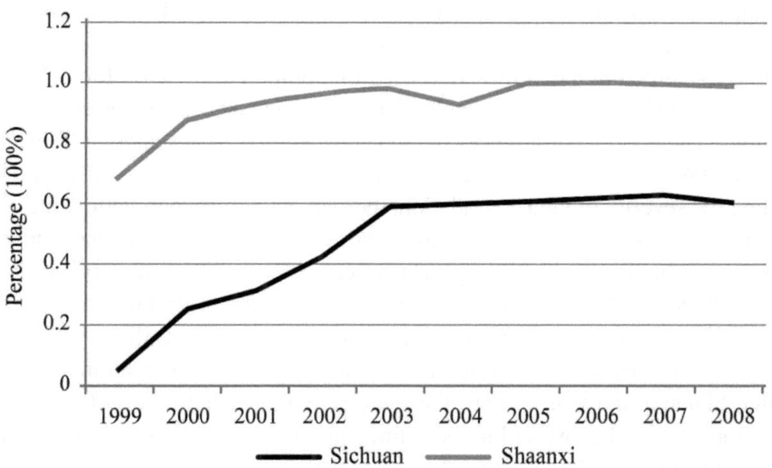


**Fig. 3.2** Program participation between 1999 and 2008 (sample households enrolled in the program out of the set target) (Source: Yin and Liu 2011)

set for 2004 and the following years (Table 3.3) (Dong et al. 2010). Concerned with food production for the whole population, the government decided to ensure that 12 million ha of cropland would be preserved nationwide.

The slowdown of the expansion of the areas of converted farmland is also depicted in Fig. 3.2 for Sichuan and Shaanxi. As Fig. 3.2 shows, participation in the GfG in Sichuan and Shaanxi grew rapidly in the first 3 years, but stabilised from 2003

(Yin and Liu 2011). In Shaanxi, a vast majority of the sample households enrolled in the program immediately after its inception and, by 2002, the quota of participating households in Shaanxi had been filled. In Sichuan, only a small portion of the households enrolled during the first 2 years, and by the beginning of 2003, participation had risen to roughly 60 % of the total amount of land the government had planned to set aside for that province. After that, no additional households were added. As noted, this was in part owing to the high level of participation reached early on, and in part owing to the government's decision to scale back the expansion of the program in light of its concern over food security.

After 2003, more emphasis was placed on caring for young forests, maintaining woods, and developing relevant industries (Cui 2009). Some emphasis was also placed on making participating households set aside more land. While the number of households that joined the program did not increase much after 2003, the annual conversion rate per household kept increasing. For example, in Sichuan province, cropland retirement began at a rate of 0.26 ha per participating family and gradually increased to around 0.33 ha (Yin and Liu 2011). By the end of 2004, the GfG had been implemented in more than 2,000 counties across 25 of China's 31 provinces, autonomous regions, and municipalities. A total of 6.1 million ha of farmland and 9.7 million ha of devastated land had already been converted to forest or grass field (Xu et al. 2004). By the end of 2006, the GfG had converted almost 9 million ha of cropland to forest/grassland and had afforested 11.7 million ha of barren land. By 2008, forestry coverage had increased by one third in the Yangtze River region, and had almost doubled in the Yellow River region compared to 1998 (Table 3.4).

National data hides the considerable variation among provinces caused by different conditions. The regional differences of GfG conversion is represented by Table 3.5 for the years 1999–2004. The differences among provinces are owing to different factors, including the severity of land degradation, the size of the rural population, the poverty of the people, and whether it is located in the watersheds of the Yangtze and Yellow Rivers. As can be seen, some provinces have seen extensive land conversion. The largest amounts during these 6 years were in Inner Mongolia (1.98 million ha), Shaanxi (1.82 million ha), Sichuan (1.53 million ha) and Gansu (1.29 million ha). Together, these four provinces account for one-third of the total area converted in all of China. Sichuan, Shaanxi and Gansu were the three provinces where the GfG started. The large amount of land converted in these three provinces is due to their importance as the watershed areas of the two largest rivers in China, the Yellow and the Yangtze, and to the extensive deforestation that occurred during the previous decades. Other large provinces, such as Yunnan, have not been reforested to the same degree because they experienced lower rates of deforestation during the previous decades.

**Table 3.4** Annual changes (percent) in forestry coverage

| | 1998 | 2000 | 2002 | 2004 | 2006 | 2008 |
|---|---|---|---|---|---|---|
| Yangtze river basin | 38.86 | 40.38 | 41.32 | 45.78 | 49.26 | 51.47 |
| Yellow river basin | 9.71 | 10.09 | 10.65 | 13.99 | 17.90 | 17.61 |

Source: Huang et al. (2009)

**Table 3.5** Converted land area of each participating province (10,000 ha)

| Province | Year | | | | | | Total |
|---|---|---|---|---|---|---|---|
| | 1999 | 2000 | 2001 | 2002 | 2003 | 2004 | |
| Anhui | | | | 26.66 | 17.34 | 2.67 | 46.67 |
| Beijing | | 0.33 | | 2.00 | 2.00 | 1.46 | 5.79 |
| Chongqing | | 4.53 | 4.53 | 29.34 | 33.34 | 10.67 | 82.41 |
| Gansu | 4.38 | 8.00 | 6.00 | 24.66 | 53.34 | 32.33 | 128.71 |
| Guangxi | | | 1.20 | 16.00 | 23.33 | 15.33 | 55.86 |
| Guizhou | | 1.80 | 2.00 | 30.66 | 34.66 | 18.66 | 87.78 |
| Hebei | | 1.47 | 1.34 | 33.34 | 48.00 | 33.67 | 117.82 |
| Heilongjiang | | 1.34 | 0.86 | 16.00 | 24.00 | 20.67 | 62.87 |
| Hainan | | | | 1.34 | 6.66 | 2.67 | 10.67 |
| Henan | | 3.00 | 3.00 | 16.00 | 25.34 | 18.66 | 66 |
| Hubei | | 1.80 | 3.33 | 22.66 | 28.00 | 12.33 | 68.12 |
| Hunan | | 2.73 | 6.67 | 40.67 | 32.00 | 12.33 | 94.40 |
| Inner Mongolia | | 13.80 | 9.33 | 64.47 | 66.27 | 44.00 | 197.87 |
| Jiangxi | | | 1.34 | 13.34 | 21.34 | 4.67 | 40.69 |
| Jilin | | 1.52 | 1.54 | 11.34 | 24.00 | 21.00 | 59.4 |
| Liaoning | | | 1.34 | 13.34 | 27.33 | 20.00 | 62.01 |
| Ningxia | | 3.06 | 3.86 | 12.00 | 18.66 | 16.66 | 54.24 |
| Qinghai | | 5.54 | 5.27 | 12.00 | 14.66 | 7.33 | 44.8 |
| Shaanxi | 28.39 | 7.40 | 10.00 | 54.00 | 56.00 | 26.67 | 182.46 |
| Shanxi | | 8.67 | 6.67 | 44.67 | 34.00 | 17.07 | 111.08 |
| Sichuan | 12.03 | 15.73 | 24.53 | 44.00 | 46.66 | 10.33 | 153.28 |
| Tianjin | | | | 0.40 | 0.40 | 0.14 | 0.94 |
| Tibet | | | | 1.34 | 1.34 | 0.67 | 3.35 |
| Xinjiang | | 0.81 | 1.40 | 17.33 | 16.00 | 7.33 | 42.87 |
| Xinjiang production and construction corps | | 0.41 | 0.80 | 5.34 | 10.66 | 4.67 | 21.88 |
| Yunnan | | 5.26 | 3.34 | 20.00 | 34.67 | 11.34 | 74.61 |
| Military[a] | | | | | 13.33 | 26.67 | 40.00 |
| Total | 44.80 | 87.21 | 98.33 | 572.87 | 713.34 | 400.00 | 1916.55 |

Source: SFA (2005c)

[a]Some of the steep-slope land are located in military areas and owned by the Military. These are not included in the provinces' aggregate data

Yin and Liu (2011) further estimated the changes in forestland and farmland in Shaanxi and Sichuan (Fig. 3.3), and found that in Shaanxi the forestland increased by 1.23 ha per household from 1999 to 2008 (from 0.57 to 1.80 ha), while the farmland decreased by only 0.47 ha from 1999 to 2008 (from 0.79 to 0.32 ha per household). On the other hand, in Sichuan the forest cover increased by 0.35 ha per household (from 0.47 to 0.83 ha), and the farmland decreased by only 0.13 ha (from 0.46 to 0.33 ha per household). In both provinces, the reduction of cropland is smaller than the increase in forestland, largely because much of the land that was

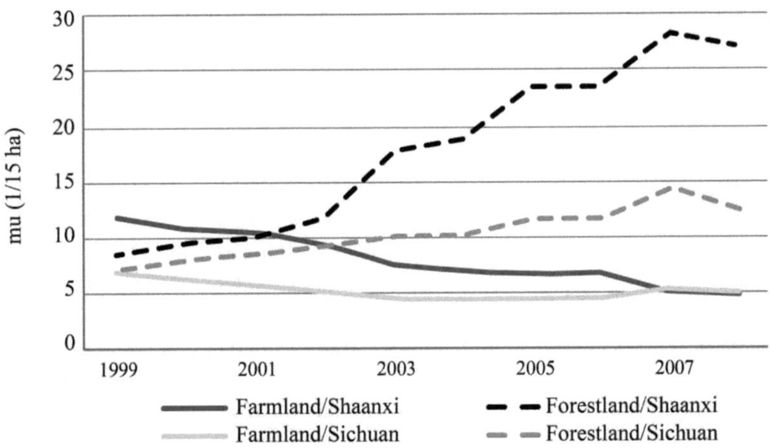


**Fig. 3.3** Cropland and forestland dynamics for sampled households (1999–2008) (Source: Yin and Liu 2011)

**Table 3.6** Area of land type in Y village (1999)

| | Land area (ha) | Farmland area (ha) |
|---|---|---|
| Total area | 375 | 166 |
| Area/family | 1.45 | 0.64 |
| Area/capita | 0.33 | 0.14 |

Source: Fukao (2000), Hori and Kojima (2008)

forested was wasteland: land that might have been farmed in the past, but had since been abandoned because of loss of soil fertility.

Most studies on the impact of the GfG have been done on smaller regions rather than whole provinces or the entire country. The following case studies describe the timeline, rate of conversion, and amount of land that was reforested through the GfG in particular areas.

## *Yulin City, Shaanxi Province*

Hori and Kojima (2008) describe the situation of Yulin City, located in the northern part of Shaanxi Province (northern part of the Loess Plateau). They looked in particular at the land use characteristics of a small village located 20 km southeast of central Mizhi County and Yulin City (Table 3.6). Yulin City was characterised by low agricultural incomes: the total amount of farmland of each household was about

0.64 ha in 1999 (Table 3.6). However, households' land consisted in several small plots of land at some distance from one another, which made cultivation more expensive and cumbersome. This was the result of the farmland allocation system at the time of the land reform in 1984, when farmland was divided evenly among village households (a pilot two-land system was first adopted in Pingdu, a county-level city in Shandong Province. By the early 1990s it had become a nation-wide policy). Since the village farmland is of different quality, it was split into parcels that were more or less uniform in terms of land fertility, distance from the villages, and other characteristics, with each household receiving one plot in each of these parcels. The result was that households ended up holding many small fields at considerable distance from one another.

Because more than half of all farmland in Yulin was experiencing desertification or was located on slope land, the government of Yulin City estimated that 68 % of farmland in Yulin should be targeted by the GfG. According to Yulin GfG regulation, the Yulin Forest Bureau called for three types of land to be subjected to the GfG: (1) farmland on slopes of more than six degrees; (2) farmland experiencing desertification; and (3) devastated land. The GfG was launched in 1999 in Yulin City, and by May 2003, 1,535,300 ha of farmland and 1,812,800 ha of devastated land had been converted to forest or grassland. By May of 2003, the farmlands that had been set aside by the GfG covered about 13.4 % of the whole farm area of Yulin City (Hori and Kojima 2008). It was expected that about two million-farmers would lose their farmland after the completion of the GfG in Yulin City.

## *Ansai County, Loess Plateau*

Ansai County, in the Loess Plateau joined the GfG program in 1999. Zhou et al. (2011) conducted a GIS analysis to investigate the gradual change of land use before and after the implementation of land conversion program. Between 1978 and 1990 the amount of croplands expanded by 21.53 %, converted from natural grassland, forests and shrubs. However, from 1990 to 1995, croplands decreased by 4.61 %, with 67.61 % of this decrease reverting to grassland (Fig. 3.4). This suggests that some of the land that farmers had converted to farmland had lost its fertility and had been abandoned. On the other hand, since the area of grassland experienced no significant rise, it is likely that farmers were forced to farm land they had previously abandoned, to address the decreasing productivity of the land under cultivation. This in turn suggests increasing population pressures and scarcity of good farmland. Only with the implementation of GfG program did the cropland area in Ansai County start to decline. Figure 3.4 shows that the most important land-cover change in the county was the transformation from agricultural land to "young afforestation land" beginning in 2000 (Zhou et al. 2011).

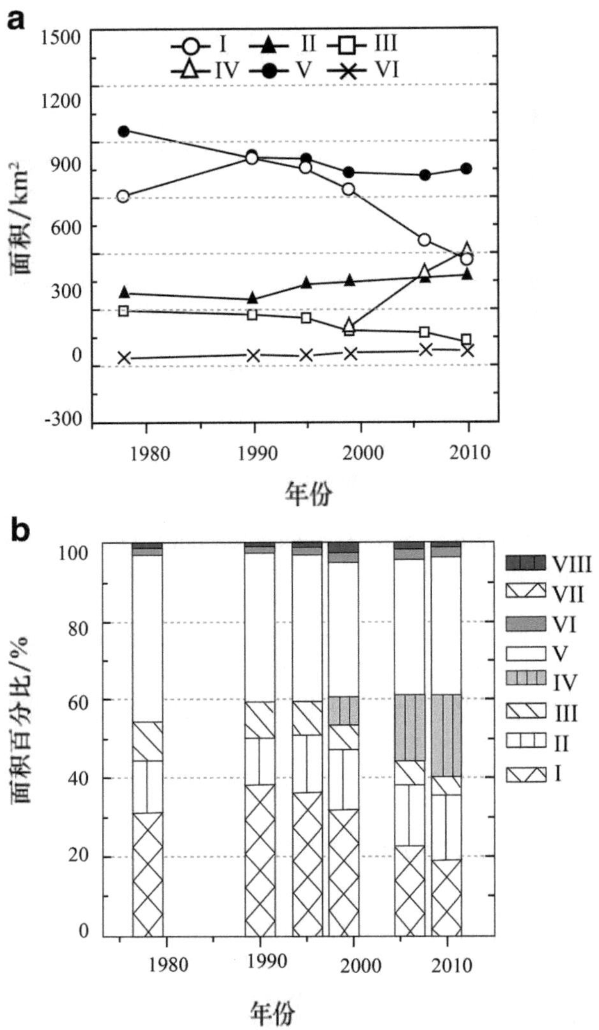


**Fig. 3.4** Area (**a**) and area percentage (**b**) of each land use /cover type in Ansai County during 1978–2010. Land use are divided into six categories (two more categories unrelated to the GfG are added in **b**) – crops (*I*), forests (*II*), shrubs (*III*), young afforestation land (*IV*), grassland (*V*), construction land (*VI*), water area (*VII*), beach land and barren land (*VIII*) (Source: Zhou et al. 2011)

## *Wuqi County, Shaanxi Province*

Wuqi County (Shaanxi Province) is an extreme example of the loss of farmland caused by the GfG. The proportion of land used for agriculture, forestry and animal husbandry changed considerably from 1997 to 2003. In 1997, 60 % of the land was used for agriculture, 34 % for forestry, and 6 % for animal husbandry. By 2003, thanks to the GfG, only 9 % was used for agriculture, while 66 % was used for forestry, and 25 % was used for animal husbandry (Ge et al. 2006). It is extreme cases such as this that increased concerns about the impact of the GfG on grain prices (discussed in Chap. 10).

## *Pingwu and Jiuzhaigou Counties, Sichuan Province*

In some villages, the hilly nature of the region and the cramped nature of the cultivated land, even when terraced, hinder the development of high-yield agriculture. This was the case in Pingwu and Jiuzhaigou Counties (Sichuan province) studied by Démurger et al. (2005). Prior to the implementation of the GfG program, three townships in Sichuan province earned a considerable part of their revenue from forestry, and to a lesser extent from agriculture. Between 2000 and 2004, nearly all the cultivated sloping land in Pingwu and Jiuzhaigou Counties (where the two townships surveyed by Démurger et al. 2005 are located) were converted, for a total of 6,666.67 and 4,400 ha respectively. However, the land conversion program placed significant pressures on subsistence communities. On the one hand, it imposed a logging ban, which reduced fuel and construction materials. On the other hand, the conversion of crops to forests reduced crop yield, and the afforestation efforts reduced grazing land. Some households converted all their land, while others were able to keep a few plots under cultivation. This was the case in the Wujiao township in particular, where each household still cultivated between 1 and 1.5 mu, after the GfG had converted much of their farmland. In addition, in a number of converted plots, there was mixed cultivation, and the planting of chestnut trees, firs, and peach trees (supported by the GfG) was combined with vegetables, allowing the villagers a subsistence level of agricultural production (Démurger et al. 2005). Nevertheless, for households living on geographically less-productive lands, land conversion was essentially land abandonment, adding to pressures for survival.

## *Jianyang City, Sichuan Province*

Zheng and Jiang (2011) assessed the positive economic impact on the farmers around Sancha Lake (Sichuan province). The lands were used for forestry and animal husbandry, and animals were often left to roam in the forests planted through the GfG, where farmers may grow fodder. Based on the official land use statistics of 1998–2002 (Anonymous 2002), Zheng and Jiang (2011) assessed the marketing value of natural resources of Jianyang city, where the Sancha Lake is located (Table 3.7). They concluded that the change in land use brought about by the GfG had great economic as well as ecological benefits. However, it is worth noting that profits per square kilometre from forestry seemed to be dropping from 1998 to 2002, while those from animal husbandry remained stable. Only in 2008 did profits increase, by half for forestry and by four to five times for animal husbandry, compared to 2002 (Table 3.7). This sudden increase raises the question of data reliability, though other possible influencing factors such as technological improvement and new market demand may have contributed to the sudden increase.

**Table 3.7** Economic profit of GfG program in Jianyang, Sichuan province

| | Area of converted land ($km^2$) | | Annual production (10,000 Yuan) | | Annual production per $km^2$ (10,000 Yuan) | |
|---|---|---|---|---|---|---|
| Year | Forestry | Animal husbandry | Forestry | Animal husbandry | Forestry | Animal husbandry |
| 1998 | 998 | 335 | 143,273 | 98,561 | 143.56 | 294.21 |
| 1999 | 992 | 341 | 141,475 | 92,339 | 142.62 | 270.79 |
| 2000 | 982 | 351 | 121,678 | 104,020 | 123.91 | 296.35 |
| 2001 | 975 | 358 | 107,754 | 110,096 | 110.52 | 307.53 |
| 2002 | 930 | 402 | 113,946 | 123,300 | 122.52 | 306.72 |
| 2008 | 1,047 | 286 | 193,400 | 430,300 | 184.72 | 1,504.55 |

Source: Zheng and Jiang (2011)

**Conclusions**

This chapter has introduced the timeline of the GfG. The GfG was introduced in 1999 in three different provinces (Sichuan, Shaanxi, Gansu). As it was a successful (and quickly became popular) program, it rapidly expanded into other provinces. In many cases, counties expanded the program beyond what was initially planned by the central government, setting aside more land than was requested, and then asking for funds. For this reason, the program expanded more rapidly than initially planned. By 2003, the GfG was blamed for creating food scarcities and increasing the price of food. Most people now argue that the increase in food prices was not actually due to the GfG (see Chap. 10), but the expansion of the program was slowed down from 2004 onward and there was a period of consolidation, or adjustment. Table 3.8 provides a timeline of the GfG's main milestones, which are discussed in this and the following chapters.

**Table 3.8** Historical milestones of GfG

| Year | Milestone |
|---|---|
| 1999 | Pilot projects started in Sichuan, Shaanxi and Gansu Provinces |
| 2000 | GfG is officially a significant part of the "Go West" campaign |
| | Pilot projects started in more provinces |
| | Relevant principles and policies are formulated |
| 2001 | GfG is officially part of tenth Five-Year Plan |
| 2002 | Nationwide official launch of GfG |
| 2003 | Peak of participation |
| 2004 | Slowdown of expansion, and structural changes |
| 2005 | Focus shifted from conversion of cropland to barren land |
| | GfG is officially part of 11th "Five-year Plan" |
| 2007 | GfG subsidy scheme renewed for another round |
| 2016 | GfG program set to end (beginning in the three provinces where the program started, later in other provinces) |

Source: Liu and Li (2010)

# Chapter 4
# Farmers' Compensation

**Abstract** This chapter looks at the level of farmers' compensation for all three plant types: economic trees, ecological trees, and grassland. Grain for Green regulations stipulate that compensation should only be paid if a large number of planted trees and grasses survive (initially 70 % in the Yellow River watershed and north China, and 85 % in the Yangtze River watershed and south China, later standardized to 75 % nation-wide). We show, however, that often compensation was also given if a smaller number of trees survived. The chapter also looks at the extent to which the funds are actually delivered to the farmers, which was a concern to the farmers when the program was introduced. Finally, the chapter reviews the total incomes of the farmers, comparing pre-Grain for Green incomes to post-Grain for Green incomes from the same land. In more cases than not, the post-Grain for Green incomes are higher than the pre-Grain for Green incomes from crop cultivation, which means that the program raised farmers' incomes. However, the Grain for Green could have converted more land with the same budget, or the same amount of land with a lower budget.

**Keywords** Payment delivery • Income • Opportunity cost • Cost-effectiveness • Ecological trees • Economic trees

## Introduction

As a Payment for Ecosystem Services (PES) program, the GfG compensates farmers for setting aside land. This chapter discusses various aspects related to that compensation. First, we look at the level of compensation paid to farmers, and the conditions the farmers need to fulfil to receive that compensation. Second, we look at the extent to which the payments due to the farmers were actually made, and the reforms that were undertaken to improve the payment system. Third, we discuss the relationship between the level of compensation and the incomes that farmers were able to obtain from the converted land.



C.O. Delang, Z. Yuan, *China's Grain for Green Program*,
DOI 10.1007/978-3-319-11505-4_4

## Overview of the Compensation Level

The total budget of the GfG has been very large, and much of it is directly used to compensate the farmers. This is unlike most other "rural development" programs, whose funds do not end up in the rural villages. By the end of 2008, the central government had invested a cumulative total of Yuan 191.8 billion in the GfG. The plans are for further investments of Yuan 240 billion, bringing the total investment to no less than Yuan 431.8 billion by 2016, when the program is set to end (Zhang 2010).

Farmers are given three kinds of subsidies for converting their land: cash, grain (which later was converted to cash compensation), and seedlings (Table 4.1). Initially, the government subsidized 150 kg of grain per year to farmers for retiring 1 mu[1] (0.07 ha) of cropland in the upper reaches of the Yangtze River, and 100 kg in the upper and middle reaches of the Yellow River. The government offered two levels of grain compensation in these two watersheds because of the soils' different fertility, which results in higher yield levels in the Yangtze River basin compared to the Yellow River basin. In 2004, the government changed the grain compensation to cash, given at Yuan 1.4 per kilo of grain, because of a shortage of grain (Cui 2009) and fear of corruption, with local officers buying cheap grain, and then billing better quality, more expensive grain to the provincial government.[2] The cash was distributed to provinces and autonomous regions where the conversion policy was adopted, and then forwarded to the participants (Li and Lu 2004). In 2004, this cash amount corresponded to the price of the grain it replaced. However, this shift from grain to cash became a problem, as this payment remained unchanged, while the price of rice increased (from an average of Yuan 1.5 per kg in 2002 to an average of Yuan 2.2 per kg in 2012). Thus, in real terms, the payment has been decreasing.

**Table 4.1** Amount of subsidies for reforestation on cropland per mu and hectare, per year

| Location of farmland | Cash (Yuan) | | Grain or cash: 1999–2003 (kg) | | Grain or cash: 2004–2006 (Yuan) | | Grain or cash: 2007–2015 (Yuan) | | Seedlings (Yuan) | |
|---|---|---|---|---|---|---|---|---|---|---|
| | Mu | Ha | Mu | Ha | Mu | Ha | Mu | Ha | Mu | Ha |
| Yangtze River watershed | 20 | 300 | 150 | 2,250 | 210 | 3,150 | 105 | 1,575 | 50 | 750 |
| Yellow River watershed | 20 | 300 | 100 | 1,500 | 140 | 2,100 | 70 | 1,050 | 50 | 750 |

Source: SFA (2000d); State Council of China (2004, 2007)

[1] *Mu* (one *mu* corresponds to 1/15 of a hectare) is the common unit of measurement for land area in China. We use hectares in this book because most authors reviewed use hectares rather than *mu*. However, all official documents mention *mu*.

[2] The provincial government was in charge of the distribution of grain subsidies, which it had to purchase from local state-owned food companies. The cost of distribution was borne by the provincial government, which was a great drain on the finances of the poorest provinces.

The farmers also receive a cash payment of Yuan 20 (for living subsidies) per mu of set-aside land per year for the duration of grain subsidies, for tending the land and miscellaneous expenses. Finally, forestry agencies supply a one-time cash subsidy to farmers for purchasing seedlings at the beginning of the conversion program. On average, the seedlings are worth approximately Yuan 50 per mu (Yuan 750 per ha) (State Council of China 2002b). In total the three types of compensation amount equal to Yuan 210 per mu (Yuan 3,150 per ha) in the middle and upper reaches of Yellow River for the first year of conversion and Yuan 160 per mu (Yuan 2,400 per ha) per year from the second year on (Table 4.1). For the Yangtze River watershed, the total amount paid was Yuan 280 per mu during the first year, and Yuan 230 from the second year on.

Compensation was conditional on the growth of the forest. Officers from the Forest Bureau verified the survival rate of the trees. To ensure that farmers planted and cared for the seedlings, only 50 % of the grain and cash subsidies were given to the farmers upon entering the program (while obviously they received 100 % of the seedlings) (Uchida et al. 2005). The farmers had to achieve a survival rate of 70–85 % of the trees to receive compensation (SFA 2001a). The remaining 50 % of the grain and cash subsidies were given when they passed the first-year inspection carried out by the local GfG implementation office (Uchida et al. 2005).[3] Farmers who did not achieve a survival rate of 70–85 % were allowed to replant the seedlings, and if the seedlings had survived when the officers from the Forestry Bureau inspected the fields again the following year, the farmers were paid retroactively (for the previous year and the present one) (State Council 2007). Because the farmer could replant every tree that had died and receive compensation retroactively, the success rate was usually officially very high, around 90–100 %.

Making compensation conditional on the survival of the seedlings is essential to guarantee that the farmers will plant the seedlings and take care of them when they are still young and need attention. However, it also has negative consequences. In some cases, farmers planted trees at a higher density than optimal, to make sure that enough survived to satisfy the government's standard of the number of seedlings per mu necessary to claim government subsidies (Yin et al. 2005).[4] If most of the trees planted survived, tree density was too high, which meant that it would take a longer time for the trees to grow and for the canopy to close, while the forest quality and ecosystem functionality were not very high. Also, forest vulnerability was increased, as were the risks of future fire and pest attacks.

Long et al. (2010) argued that forest management activities following tree planting, such as competition control and thinning, were poorly incorporated into the program, because many rural workers migrated to urban areas. Most of those who remain in the rural areas are old people, sick or disabled men or women, and children, who are unable to manage large forest areas.

---

[3] Xu and Cao found (2002) that this advance payment system had not been adopted in some areas.

[4] According to Trac et al. (2007), in 2003, the average density for the monitored counties was reported as 148 seedlings per mu (2,220 seedlings per hectare or about one seedling for every 4.5 $m^2$).

## Payment Delivery to Farmers

A number of researchers have addressed the question of whether the payments were actually delivered to the farmers. Zuo et al. (2003) found several cases during the pilot phase where full compensation did not reach participating farmers. Similarly, Xu and Cao (2001) found that in a group of 1,026 households, fully 49.5 % had received only partial compensation, 8.5 % had received only grain and 17.6 % had received no compensation at the time of the survey.

Bennett (2008)[5] also looked at whether the compensation was actually delivered to the farmers and concluded that there was some evidence of significant shortfalls in subsidies actually delivered (Table 4.2). In some cases, shortfalls may have been the result of plots that have been converted but have not yet been fully certified

**Table 4.2** Average shortfalls in grain and cash compensation, 2002

| | | | Grain (kg/ha)[a] | | Cash (Yuan/ha) | | Total shortfall (Yuan/ha)[b] |
|---|---|---|---|---|---|---|---|
| Province | County | Township | GfG standard | Actual delivery | GfG standard | Actual delivery | |
| Shaanxi (n=103) | Yanchuan | Yanshuiguan | 1,500 | 506 | 300 | 25 | 1,269 |
| | | Majiahe | | 466 | | 59 | 1,276 |
| | | Yuju | | 94 | | 8 | 1,698 |
| | Liquan | Yanxia | | 1,074 | | 112 | 614 |
| | | Jianling | | 1,500 | | 48 | 252 |
| | | Chigen | | 1,471 | | 78 | 251 |
| Gansu (n=85) | Jingning | Zhiping | 1,500 | 574 | 300 | 104 | 1,122 |
| | | Gangou | | 957 | | 137 | 707 |
| | | Lingzhi | | 1,170 | | 201 | 429 |
| | Linxia | Zhangzigou | | 499 | | 86 | 1,215 |
| | | Tiezhai | | 0 | | 5 | 1,795 |
| | | Hexi | | 588 | | 36 | 1,176 |
| Sichuan (n=76) | Chaotian | Datan | 2,250 | 1,849 | 300 | 87 | 614 |
| | | Zhongzi | | 2,050 | | 0 | 500 |
| | | Shahe | | 2,177 | | 39 | 334 |
| | Li | Shangmeng | | 2,160 | | 107 | 284 |
| | | Puxi | | 2,250 | | 231 | 69 |
| | | Guergou | | 618 | | 50 | 1,882 |
| Average: | | | | 856 | | 70 | 1,021 |

Source: Bennett (2008)

[a]This is a sum of corn, wheat, white and paddy rice, and wheat flour subsidies. Both white rice and wheat flour were converted to unhusked weight equivalents at a factor of 1:1.4

[b]This values grain at the national price of Yuan 1/kg

[5]Bennett (2008) rests his analysis upon a 2003 household and village-level survey conducted by the Center for Chinese Agricultural Policy, Chinese Academy of Sciences, and another survey in Hunan Province.

under the GfG. However, Bennett (2008) argued that, in general, these shortfalls did not appear to be the result of program lag time, since the maximum average shortfall (Yuan 1,507.5/ha) was in Yanchuan county in Shaanxi Province, where implementation generally started earliest in the sample, while the minimum (Yuan 480/ha) was found in Chaotian county in Sichuan Province, where implementation generally occurred latest. Bennett (2008) argued that these shortfalls could have been due to different reasons, not all related to poor program budgeting. In some cases, shortfalls may have been the result of plots that were converted but had not yet been fully certified under the GfG. In other cases, shortfalls may have been due to deductions by village governments to either pay laborers to plant trees on the farmer's converted land, to pay for other administrative costs, or to pay back-taxes owed by the farmer (Zuo et al. 2003; Xu and Cao 2001). Also, program coordination, inspection and compensation delivery for millions of plots is burdensome and costly for local governments, yet the GfG plan dictates that local governments bear their own implementation costs (Bennett 2008).[6] Thus, the delayed payments were partially a result of the fast expansion of the program, which created even greater administrative needs, and shortfalls in required administrative funds. These, in turn, led to problems in implementation and subsidy delivery (Bennett 2008). For example, Bennett and Xu (2005) pointed out that "in a township in a key project county in Shaanxi Province, half of the participating plots were not inspected and compensated on time. In another township in the same county, many participating plots had yet to be inspected even 3 years after they had joined the GfG. Though the county government recruited 30 additional staff to deal with these problems, manpower was still far short of that required to inspect some 67,000 ha of converted land" (Bennett and Xu 2005: 12).

In 2004, after the central government became aware that local authorities were siphoning off the compensation they should have paid to the farmers, the method of compensation was changed (Delang and Wang 2013). From that point, the money was paid through the Rural Credit Cooperative and recorded in a passbook, so the farmers could verify how much they received. In some cases, farmers realized that not all money due to them was paid, and sued the Forest Bureau.[7] By transferring funds directly into the bank account of the farmer, the government could ensure that the farmer received the funds. While this was done to address the risk of corruption, this problem does not seem to have completely disappeared. Du (2012) argued that abuse of power by forestry officers increased because of the GfG.

Payments through the Rural Credit Cooperative increased the transparency of the system (FDOGX 2006). Another factor that increased the transparency of the GfG was the increasing use of the Internet. Since 2000, the central government has been encouraging all levels of government (at the national, provincial and county levels) and

[6] Since 2002 the central government has allocated some administrative fees to provincial governments for GfG implementation, but these have been insufficient and a significant percentage are often diverted by higher levels of government before reaching the townships.

[7] The legal cases have increased since 2006, as the farmers understand better how the GfG payments work.

government departments to make official websites in order to keep the public informed and announce policies. In 2007, this was codified as part of the obligations of all levels of government and all departments. By the end of 2007, more than 30,000 websites by all provinces and national-level government departments, 98.5 % of city-level authorities, and 85 % of county-level governments had established their own websites to announce their most important policies, including those concerning forestry (Yuan 2010; State Council of China 2007; Wang et al. 2005). Almost all villages in China have access to the Internet through mobile phones and farmers unable to access the Internet can obtain information from children, relatives, or friends. By learning about the policies not only from forestry officers, but also directly from the MOF or other government websites, farmers can fully inform themselves of their rights and obligations (Delang and Wang 2013).

## Total Incomes of Farmers

Central to the realization of the long-term goals of the GfG is whether it generates the right financial incentives for the participants. When the program was introduced, the subsidies needed to at least offset the participants' opportunity cost of the set-aside land. Once the subsidies end, the farmers should earn from their new timber forests, orchards or pastures (in addition to off-farm incomes they may now be earning) more than they would from pre-GfG land uses, or they would revert their land back to pre-GfG land uses. Since, in most cases, subsidies cannot be paid indefinitely, post-program land use decisions of participating farmers have been one of the biggest concerns in conservation set-aside programs elsewhere (Cooper and Osborn 1998).

A number of researchers have looked at how the subsidies compare to the pre-GfG incomes from farming land. Uchida et al. (2005) found that around 24 % of their sample households in Ningxia Province, and 77 % of their sample households in Guizhou Province, received payments which corresponded to less than the pre-program net revenue from the plots. However, their study was based on a 2,000 dataset, collected only a few months after the program had been implemented, and when the full economic benefits of land conversion could not yet have been realized. The same dataset informed the study published by Uchida et al. (2007) and the same concerns can therefore be raised. Uchida et al. (2007) used propensity scoring matching to evaluate the social and economic impacts of the program. Overall, they found evidence of a significant negative impact on cropping income. However, they also looked at other sources of income, and found a significant positive impact on husbandry income and inventories, and a significant positive impact on productive and housing assets (Xu et al. 2010). Altogether, they estimated the impact on total household per capita income to be small and statistically insignificant.

Bennett (2008) looked at the same issue using a 2003 household and village-level survey conducted by the Center for Chinese Agricultural Policy, Chinese Academy of Sciences, and an additional survey in Hunan Province. He found that for many participants, GfG compensation standards were significantly below the 1999 (pre-GfG) net incomes of the enrolled plots. The results are summarized in Table 4.3, which compares annual net income of enrolled plots in 1999 (i.e. before

**Table 4.3** Participant 1999 net income from enrolled land versus GfG compensation standards (2003 survey data)

| | Net losing households | | | Net gaining households | | | All participants | | |
|---|---|---|---|---|---|---|---|---|---|
| | Shaanxi (n=103) | Gansu (n=85) | Sichuan (n=76) | Shaanxi (n=103) | Gansu (n=85) | Sichuan (n=76) | Shaanxi (n=103) | Gansu (n=85) | Sichuan (n=76) |
| Number of households | 7 | 42 | 23 | 96 | 43 | 53 | 103 | 85 | 76 |
| 1999 average net income from enrolled land (Yuan/ha) | 4,833 | 3,485 | 5,371 | 181 | 940 | 1,031 | 507 | 2,026 | 2,457 |
| Total converted land (ha) | 5.13 | 8.52 | 7.47 | 68.11 | 11.44 | 15.26 | 73.24 | 19.97 | 22.73 |
| Average difference between GfG standard and 1999 net income (Yuan/ha)[a] | –3,033 | –1,685 | –2,821 | 1,619 | 860 | 1,519 | 1,293 | –226 | 93 |

Source: Bennett (2008)

[a]Subsidy grain was converted to cash based on the national market price of Yuan 1/kg

they were enrolled) with the amount of subsidies that they should have received for these plots in 2002 according to the program standards (in reality payments were below that level in 2002, as mentioned in Chap. 3). In Gansu, almost 50 % of the participants in the sample experienced a shortfall, which averaged 8 % of the 1999 (pre-GfG) incomes. In Sichuan, about 29 % of the participants in the sample experienced a drop of income, with shortfalls for these households averaging 11 % of the 1999 net income. In Shaanxi 7 % of participants experienced an average shortfall of almost 33 % of the average 1999 net household income. Furthermore, many households reported that the 1999 harvest was poor, which means that the losses experienced by the farmers would on average be higher.

Xu et al. (2010) looked at the restructuring of agricultural production initiated to the GfG in Shaanxi, Gansu and Sichuan provinces, and found that the GfG has indeed induced a restructuring of agricultural production, whereby participants shifted relatively more of their inputs out of cropping and into husbandry. In Shaanxi Province, growth rates for cropping income were 35 % for non-participants compared to only 12 % for participants (including subsidies received). In Gansu, these were −26 % and −32 %, respectively, and in Sichuan cropping income declined by 30 % for both groups (Xu et al. 2010). Table 4.4 presents the 1999 and 2002 components of total income for participant and non-participant households, by province.[8] Conversely, growth rates for husbandry were higher for participants than for non-participants. In Shaanxi, average per capita household husbandry income for participants increased more than ten-fold, compared to only 175 % for non-participants (Xu et al. 2010). In Gansu, participants' husbandry income grew by 1,744 %, compared to 586 % for non-participants, and in Sichuan these numbers were 845 % and 514 %, respectively. Differences in change of total income between participants and non-participants are less systematic across regions. In Shaanxi, total income (including subsidies received) increased by 41 % and 42 % for participants and non-participants, respectively. For Gansu these numbers were 2.3 % and 12 %, respectively, and for Sichuan they were 26 % and 17 %, respectively (Xu et al. 2010).

The analysis of Xu et al. (2010) was done using the results of a 2003 survey, only 2–4 years after the program was implemented in the villages surveyed. It is likely that, as time went by, the incomes from the program's land use changes (including off-farm work they may have engaged in) would have increased even further. Furthermore, Xu et al. (2010) stated that "the GfG subsidy is calculated as the subsidy received by the household for 2002", while Bennett (2008) (using the same

[8] Xu et al. (2010) considered cropping income to consist in total crop production valued at average village market price, net of materials and hired labor costs. Husbandry income includes both sales income and own consumption, valued at market prices. Off-farm income includes all nonagricultural production activities, comprised mainly of sideline activities and wage labor income. Income from sideline activities is net of production costs and other business related expenditures. Wage income includes both cash and in-kind income, valued at market prices. Other income consists of aquaculture, rental and interest income, gifts, pension income, and government subsidies and transfer payments. The GfG subsidy is calculated as the subsidy received by the household for 2002 (Xu et al. 2010).

**Table 4.4** Per capita net incomes of participant and non-participant households (1999 and 2002)

| | Nonparticipant households | | | | Participant households | | | |
|---|---|---|---|---|---|---|---|---|
| | 1999 | | 2002 | | 1999 | | 2002 | |
| Income component[a] | Mean | St. Dev. | Mean | St. Dev. | Mean | St. Dev. | Mean | St. Dev. |
| *Shaanxi* | | | | | | | | |
| Total without subsidy | 940 | 777 | 1,335 | 930 | 986 | 1,077 | 1,325 | 1,874 |
| Total with subsidy received | – | – | – | – | – | – | 1,394 | 1,877 |
| Cropping without subsidy | 465 | 521 | 626 | 429 | 420 | 672 | 401 | 622 |
| Cropping with subsidy received | – | – | – | – | – | – | 470 | 628 |
| Husbandry | 6 | 23 | 17 | 63 | 18 | 78 | 208 | 916 |
| Off-farm | 388 | 623 | 590 | 947 | 401 | 554 | 525 | 680 |
| Other | 82 | 233 | 101 | 234 | 147 | 686 | 191 | 826 |
| *Gansu* | | | | | | | | |
| Total without subsidy | 1,803 | 1,681 | 2,021 | 1,741 | 1,287 | 980 | 1,287 | 942 |
| Total with subsidy received | – | – | – | – | – | – | 1,317 | 942 |
| Cropping without subsidy | 484 | 350 | 360 | 246 | 589 | 523 | 370 | 320 |
| Cropping with subsidy received | – | – | – | – | – | – | 399 | 345 |
| Husbandry | 17 | 53 | 119 | 220 | 6 | 30 | 113 | 222 |
| Off-farm | 1,192 | 1,570 | 1,346 | 1,624 | 633 | 679 | 681 | 647 |
| Other | 110 | 515 | 196 | 541 | 59 | 204 | 124 | 393 |
| *Sichuan* | | | | | | | | |
| Total without subsidy | 1,419 | 1,425 | 1,654 | 1,271 | 1,635 | 1,195 | 1,961 | 1,524 |
| Total with subsidy received | – | – | – | – | – | – | 2,067 | 1,514 |
| Cropping without subsidy | 721 | 938 | 506 | 633 | 829 | 931 | 472 | 590 |
| Cropping with subsidy received | | | | | | | 577 | 583 |
| Husbandry | 33 | 42 | 202 | 200 | 49 | 75 | 459 | 1,187 |
| Off-Farm | 543 | 953 | 714 | 987 | 674 | 897 | 869 | 971 |
| Other | 122 | 295 | 232 | 476 | 83 | 251 | 161 | 375 |

Source: Data from Table 6, Xu et al. (2004)
Source: Xu et al. (2010)
[a]All units are in 1999 Yuan, adjusted using the Rural Consumer Price Index

dataset) states that in 2002 farmers received much lower payments than the amount they were due according to government regulations. If Xu et al. (2010) had used full GfG payments, which farmers tend to receive since 2004 (see Chap. 3), their incomes after joining the GfG would have been even higher.

We can conclude that existing studies (all carried out within the first 4 years of the program) show that the incomes of some participants may have dropped if animal husbandry and off-farm work are not included, but may have increased if animal husbandry and off-farm incomes are taken into consideration. It is not surprising that not all farmers have experienced a similar increase – or drop – in income. Since payments are uniform within each of the two regions (the Yangtze and the Yellow rivers watersheds), farmers who have converted very poor land are likely to have experienced an increase in income, while farmers who have converted more fertile land may have experienced a drop of income. As both on-farm and off-farm incomes are likely to have increased further over the last 10 years, we may conclude that participation in the GfG has generally been financially rewarding. On the other hand, we should also consider that 1999 was a year of poor harvest due to serious drought in the surveyed regions, which means that the 1999 cropping income was below the cropping income of an average year. This implies that using the 1999 cropping income as the opportunity cost for program participation is more likely to underestimate participant farmers' real opportunity costs. Further, the fact that government subsidies in 2002 were lower than the 1999 cropping income for a significant share of participant farmers was indeed a serious issue (Xu et al. 2010).

One also has to recognize that these payments are compensation for setting aside the land, and little work is necessary after the initial planting, weeding, and caring for the seedlings.[9] Thus, farmers are free to engage in other income-generating activities, either on the farm or elsewhere. Thus, income from GfG-subsidies does not need to be as high as income from farming for the farmers to benefit from converting their land. Furthermore, many risk-averse households might prefer a lower guaranteed subsidy over a higher but highly variable farming income. It would also be useful to compare incomes per person-day of farm work, rather than incomes per hectare. Incomes per person-day are likely to have sharply increased.

## Compensation and Opportunity Cost

An important question is whether the level of compensation is suitable. Compensation that is too low increases the costs of the participating farmers, and may compromise the ability of the poorest farmers to convert their steeply sloped land, thus weakening the poverty alleviation goal, among other problems. Excessive compensation means that less land can be converted, given the limited budget. Ideally, plots with the lowest opportunity cost of the land should be converted, and households should be compensated the same amount as their loss.

[9] This of course raises the question of what will happen once the subsidies end, in 2015–2016. Whether these higher incomes will continue once the subsidies end, will be discussed in Chap. 10, which deals with the sustainability of the program.

**Table 4.5** Comparison of yields and slopes from case studies in China's Grain for Green (GFG) Program, 2000

| Counties in case study | Average yield before program (kg/ha) | | Total area set aside (ha) | Proportion of land with slope 15° or greater (percent) | | Grain payment received per hectare (kg) |
|---|---|---|---|---|---|---|
| | Plots set aside under GfG | Plots not set aside under GfG | | Cropland set aside under GfG | Cropland not set aside under GfG | |
| Dingxi, Gansu | 1,369 | 2,220 | 2,000[a] | 83 | 45 | 1,500 |
| Zouzi, Inner Mongolia | 1,125 | – | 9,367[b] | 16 | 33 | 1,500 |
| Pengyang, Ningxia | 1,464 | 2,076 | 5,080 | 93 | 72 | 1,500 |
| Heqing, Yunnan | – | – | 1,000 | 96 | 91 | 2,250 |
| Dafang, Guizhou | 2,329 | 2,731 | 1,333 | 98 | 69 | 2,250 |
| Tianquan, Sichuan | 3,106 | 8,646 | 4,600 | 86 | 65 | 2,250 |

Source: Uchida et al. (2005) (Adapted from Xu and Cao 2002)
[a]Data from 2001
[b]Includes areas of afforested barren hills

Uchida et al. (2005) looked at the productivity of the land, and concluded that plots that have lower opportunity cost were usually selected for the program, making the program rather cost-effective. However, they also pointed out that within the group of participating and non-participating plots, there is substantial heterogeneity (Table 4.5): nearly 40 % of the plots in their sample had yields that were usually lower than the compensation rate (1,500 kg per hectare per year in the Yellow River Basin, and 2,250 kg in the Yangtze River Basin). The owners of the lower yielding plots were in some sense being over-compensated. In Ningxia County 15 % of the program plots had higher net revenue than the compensation level (Yuan 140 per mu), while nearly 70 % of the non-program plots had lower net revenues than this level. On the other hand, in Guizhou, 40 % of the program plots had higher net revenue than the compensation level (Yuan 210 per mu), while nearly 30 % of the non-program plots had lower net revenue than this level. Despite the fact that program plots had lower net revenues on average than non-program ones, targeting was far from perfect. Having such a large portion of the plots either above or below the compensation rate is an indicator of poor efficiency. Better targeting could have reduced the cost to the government and increased the profits of participating farmers, by including non-program plots that had lower net revenues instead of the relatively more profitable program plots (Uchida et al. 2005).[10]

[10]Uchida et al. (2005) do not have precise information as to why the excluded plots were not selected for the program, but believe that it may have been partially due to some program selection strategies adopted by local officials. For example, in some regions local officials required the plots to be contiguous to each other or to be located along a road, to minimize implementation costs.

**Table 4.6** Actual compensation vs. compensation based on net revenue for total area under the GfG in Ningxia and Guizhou, 2000

| | Ningxia | Guizhou |
|---|---|---|
| | (Yuan) | |
| Actual compensation for program plots (A)[a] | 137,942 | 21,364 |
| Amount of over-compensation (B)[b] | −75,557 | −1,994 |
| Amount of under-compensation (C)[b] | 24,063 | 6,603 |
| Compensation based on net revenue (D=A+B+C) | 86,448 | 25,973 |
| | (percent) | |
| (A)/(C)×100 | 160 | 82 |
| | Ningxia | Guizhou |
| Actual compensation for program plots (yuan) (A)[a] | 137,942 | 21,364 |
| Amount of over-compensation (yuan) (B)[b] | −75,557 | −1,994 |
| Amount of under-compensation (yuan) (C)[b] | 24,063 | 6,603 |
| Compensation based on net revenue (D=A+B+C) | 86,448 | 25,973 |
| (A)/(C)×100 (%) | 160 % | 82 % |

Source: Uchida et al. (2005)
Data: Authors' survey
[a]To calculate the actual compensation this study assumes that the farm households in the survey were fully compensated for their program plots
[b]The amounts of over-compensation and under-compensation were derived by taking the difference between the estimated net revenue and compensation per mu for each plot and then multiplying by the plot area

The data also illustrate how the degree of over-compensation varies across the study areas and reveal the potential to improve the cost-effectiveness of China's Grain for Green program. To show this, Uchida et al. (2005) compared the program payments and the level of compensation needed to compensate the household for its lost net revenue (Table 4.6).[11] In Ningxia, 84 % of the program plots had payments (Yuan 140 per mu) that were higher than the net revenue that the plot earned during the year before it was entered into the program. The average gap between the plots' payment and their net revenue exceeded Yuan 80, a level that is nearly 58 % of the compensation level (Uchida et al. 2005). If officials had compensated farmers at levels equalling the plots' pre-program net revenues, they could have reduced expenditures by 60 %. In contrast, in Guizhou 60 % of the program plots had payments (Yuan 210 per mu) that were higher than the plots' net revenue, with an average overpayment of about 39 %. Meanwhile, the amount of under-compensation exceeds that of over-compensation, resulting in net under-compensation. Officials would have had to increase expenditures by 18 % to eliminate the under-compensation (Uchida et al. 2005).

Uchida et al. (2005) argues that targeting based on these rules is likely to lead to selection of plots that do not have high slopes. While implementation costs cannot be ignored, they need to be weighed against the benefit of selecting highly-sloped plots.

[11]The analysis requires two new variables: *over-compensation*, generated by subtracting the actual payment from the plot's net revenue when actual payment is greater than net revenue, and *under-compensation* when actual payment is smaller than net revenue (Uchida et al. 2005).

On a household basis, 76 % of participating households in Ningxia and 23 % in Guizhou received payments that exceeded the net revenue that they had made on the plots the year before (Uchida et al. 2005). For a majority of the program plots, farmers received more in payments after entering the GfG program than they had received from planting crops. From the household's point of view, the GfG must have been considered a lucrative program. If the results of Uchida et al.'s (2005) sample was indicative of the situation across China, their findings implied that China would have gained by reallocating resources across regions and among households. For example, Ningxia could have improved its cost-effectiveness performance considerably by targeting those plots with higher slopes and lower opportunity costs. It should be recognized, however, that perfect targeting typically cannot be achieved in practice since there are transaction costs involved in collecting information (Uchida et al. 2005). In addition, as Uchida et al. (2005) noted, one of the main problems arising from a bidding process for contracts, such as that practiced with the Conservation Reserve Program (CRP) in the US, was that the bidding process itself affected the rental rates. Hence, Uchida et al. (2005) believed that the Grain for Green program could have benefited by adopting a more flexible payment schedule, with payments better tailored to the opportunity costs of the land, or the characteristics of plots, but not necessarily a bidding process. Indeed, Uchida et al. (2005) argued that the bidding system was not a realistic option in rural China where the administrative costs to set up such a mechanism would have been prohibitive.

On the other hand, Xu et al. (2010) argued that the use of market-based voluntary mechanisms of participation is key to the efficiency gains promised by payment for environmental services programs over traditional command-and-control approaches. In the case of the GfG, since no bidding mechanism exists to optimally match payer benefits with participant costs, participation should, at minimum, be voluntary. This would have improved cost-effectiveness, by ensuring that households with the lowest opportunity costs participated, while minimizing the possibility that program participation was having negative welfare effects on some participants (Xu et al. 2010).

Before drawing final conclusions about cost-effectiveness, however, we also need to take into consideration the environmental benefits. Uchida et al. (2005) did this by accounting for both opportunity costs and environmental benefits for each group of plots, categorized by their slope.[12] Uchida et al. (2005) found that all of the plots entering the program in Guizhou had high slopes, implying that in that province the program largely targeted plots that gave maximum environmental benefits. At the same time, some plots had high net revenues before entering the program. These plots could have been replaced by those having high slope and lower net revenue. In contrast, in Ningxia the costs and benefits were unsystematically dispersed. For example, 11 set-aside plots in the sample had no slope and high net revenue, while 45 set-aside plots had moderate slopes and low to high net revenues. Based on the observation that there wee a number of plots with higher slopes and lower net revenue per mu, the figures suggest that, from the cost-effectiveness point

[12]The survey respondents classified each of their plots in three levels: those with steep slopes (over 25°), moderate slopes (15–25°) and others (less steep and flat).

of view, the site selection was not performed well in Ningxia. Ningxia could have improved its cost-effectiveness performance considerably by targeting those plots with higher slopes and lower opportunity costs (Uchida et al. 2005).

Uchida et al. (2005) concluded that China's government can improve cost-effectiveness in two ways. First, the program can decrease costs and avoid hurting farmers by reducing the cases of over-compensation and increasing the compensation for (or removing from the program) the plots that are being under-compensated. In a similar way that is done in the CRP in the United States (Babcock et al. 1996), this can be accomplished by changing the compensation schedule from a uniform rate to a more flexible payment schedule that is based on the actual opportunity costs and environmental benefits of each plot. Second, the program can maximize its cost-effectiveness by weighing both the opportunity cost and environmental benefit of each plot, and target as precisely as possible those sites that have low opportunity costs and high environmental benefits (Uchida et al. 2005).

## Ecological and Economic Trees

The GfG scheme converts croplands and wasteland into two kinds of forests: ecological or economic forests. Ecological forests are defined as timber-producing forests, while economic forests are orchards or plantations with trees of medicinal value, or other trees providing non-timber forest products that may be sold by the farmers (SFA 2001c) (Chap. 6 discusses the characteristics of the two kinds of forests in more detail).

According to the GfG regulations, farmers received grain subsidies for 8 years if they converted land to ecological forest, 5 years for economic forest and 2 years if they converted land to grassland (Yin et al. 2005). Hence, the GfG was set to expire in a maximum of 8 years after it was first introduced – between 2007 and 2012 in most areas. Because of the fear that the forests did not yet generate sufficiently high incomes to compete with farmland (for example, fruit trees need a number of years before producing an income), and that farmers would cut the trees and revert the land back to pre-conversion land use, the program was extended for one additional period in 2007. That is, farmers would be compensated another 8 years for ecological trees, 5 years for economic trees, and 2 years for grassland. However, compensation was halved (grain subsidy dropped from Yuan 3,150 to Yuan 1,575 per ha of converted land in South China and the Yangtze River basin, and from Yuan 2,100 to Yuan 1,050 per ha of converted land in the Yellow River basin [Table 4.1]). As mentioned above, until 2004 the farmers received grain, which on average corresponded or exceeded the value of the agricultural produce they were able to grow on the land they had set aside. In 2004, this grain compensation was replaced with cash, which corresponded to the price of the grain. By 2007 the cash compensation was already below the potential cash incomes from the land, because the price of grain had increased. In 2008, as the government halved that (already low) cash compensation, the incomes to farmers were further drained, and the opportunity costs increased. Local governments had the option to increase the financial compensation, but most did not (Delang and Wang 2013).

Halving the compensation paid to the farmers does not mean that government expenditure fell. While the compensation paid directly to the farmers was approximately halved after 2007, a similar amount was used to strengthen the overall results of the program through government investment in different areas. One area the government invested in was to improve the quality of the grain fields. This was done through water conservancy work, such as levelling the ground and constructing irrigation ditches to improve the efficiency of water utilization, so as to improve the productivity of flatland and reduce the dependence of farmers on slope land (Li 2007). The government also invested in research to select the trees used for reforestation and to improve seedlings. It supported agricultural extension work to improve the quality of the soil (and seedlings) and to provide training to plant and manage ecological and economic trees. Finally, the government invested through the GfG to increase the availability of alternative energy sources to replace firewood, since people's needs for wood to cook and heat the house was one of the causes of forestland degradation.[13] The government addressed the energy needs of villagers in a targeted way by promoting energy-saving stoves, biogas digesters and liquefied gas, or through projects such as establishing firewood forests and setting up small hydropower plants. Lastly, when villages were located in areas that were ecologically fragile, funds were used to resettle villages to a more desirable location. Hence, overall subsidy levels have not dropped but have increased somewhat (Li 2007; Delang and Wang 2013).

Table 4.7 breaks down in different categories the available data on levels of government expenditure from 2000 to 2010. The data are only indicative, as total government investment in the GfG was much larger than the figures reported in the table. However, the table may give the reader an idea of the different kinds of expenditures, and the proportion of each category in relation to the others during the period under consideration.

As mentioned, in 2007 the GfG was extended for another 8 years. By 2012, 60 % of the land that had been converted was receiving subsidies from the second phase subsidy scheme (SFA 2013c).[14] In 2012, in the counties sampled, the areas of converted cropland that were receiving their first round and second round of subsidies was 30.02 % and 64.3 %, respectively, while 5.68 % of the land set-aside through the GfG was no longer subsidized because the subsidy period for that land had already ended (SFA 2013c).

[13] For example, in Yunnan Province, at the end of 1998, 76 % of administrative villages used brushwood as an energy resource (Meng et al. 2000: 27–32), while in Lijiang Prefecture, Nujiang Prefecture, and Diqing Prefecture in the northwest of China, 40.3 % of peasant households still did not have a power line, and 100 % of the energy used came from burning brushwood (Meng et al. 2000: 27–32). If the resource models in these three prefectures had not changed and they had continued cutting wood, as much as 690,000 ha of woodland could have been lost every year in these three prefectures alone, and forest resources could have been preserved for only 57 years, after which there would have been no wood to burn (Meng et al. 2000: 27–32) (Delang and Wang 2013).

[14] First phase refers to the subsidy scheme from 1999 to 2007, second phase refers to the subsidy scheme from 2008 to 2015.

**Table 4.7** Statistics of GfG subsidies

| Year | Cash subsidy (billion Yuan) | Government investment (billion Yuan) | | | | | Subsidized household (million) |
|---|---|---|---|---|---|---|---|
| | | Grain subsidy | Seedling | Technology input | Others | Total | |
| 2000 | | | 0.333 | | | 1.541 | |
| 2001 | 0.35 | 2.036 | 0.737 | 0.012 | 0.441 | 3.214 | |
| 2002 | 0.458 | 6.308 | 3.307 | 0.032 | 1.446 | 11.061 | 10.31 |
| 2003 | 2.818 | 942.8[a] | 5.481 | | | 22.599 | 18.85 |
| 2004 | | 16.824 | 2.981 | | | 23.574 | 23.28 |
| 2005 | 2.533 | 284.18[a] | | | | 26.812 | |
| 2006 | 2.574 | 22.390 | | | | 25.810 | >28 |
| 2007 | 2.703 | 22.671 | | | | 23.514 | approx. 30 |
| 2008 | 3.184 | 21.448 | | | | | approx. 30 |
| 2009 | 3.102 | 20.651 | | | | | 28.38 |
| 2010 | | | 1.328 | | 10.572 | 32.205 | 27.53 |

Source: Ke (2007), SFA (2002–2011)
[a]The unit for these 2 years is kg

**Conclusions**

This chapter has reviewed the level of compensation paid to the farmers and how this compensation changed over time. In many cases, the program paid more to the farmers than the income they received from farming. This means that either the program could have saved money by paying the opportunity cost of the land, or more farmland could have been converted using the same amount of funding. It is worth noting that farmers may still benefit from joining the GfG even if the payments they receive are below the opportunity cost of the land, because after joining the GfG they are free to engage in other income-generating activities, either in the village or elsewhere.

One of the advantages of the GfG, in contrast to other programs, is that most of the money allocated to the GfG by the central and provincial governments directly ends up in farmers' pockets. This, of course, contributes to the popularity of the program and to the fact that in many places more farmers wanted to join the program than were eventually allowed to, because of budgetary constraints (Chap. 7). The next chapter discusses GfG's process of land selection.

# Chapter 5
# Land Selection

**Abstract** This chapter reviews land selection. According to program guidelines, slope land or unproductive land prone to soil erosion should have been converted first. Evidence shows that although this has happened in many cases, it was not universal. In some cases, productive land with low slope was also converted. Researchers argued that this might have been because program managers preferred to convert large tracks of adjacent land, and flat productive land might have been found between steeply sloped land.

**Keywords** Farmland • Soil erosion • Slope land • Degraded land • Land productivity

## Introduction

The Grain for Green is primarily designed to reduce the amount of steep sloped farm land. Farming steep slopes has been illegal for many decades, and throughout the decades many laws have been implemented to stop the practice.[1] The laws, however, were usually ignored because poor farmers needed to supplement their income through the cultivation of marginal land. From an environmental perspective, the GfG tackled the same problem and had similar objectives, but instead of state power it used financial incentives and market mechanisms, which proved to be more effective (Delang and Wang 2013).[2]

The GfG attempted to convert land of limited suitability for agricultural production. This was defined primarily as slope lands that were subject to excessive soil erosion (Xu et al. 2005). Xu and Cao (2002) and Uchida et al. (2007) noted that, in the GfG program, slope land was defined by a slope greater than 15° in the northwest region and greater than 25° in the southwest region. Hori and Kojima (2008) added that the GfG targets not only slope land, but also three other types of land:

[1] For example the "Law of the People's Republic of China on Water and Soil Conservation" (中华人民共和国水土保持法) of 1982.

[2] In this sense, the GfG was revolutionary in the Chinese context; it was the first time that the market mechanism was employed instead of command-and-control approaches.



C.O. Delang, Z. Yuan, *China's Grain for Green Program*,
DOI 10.1007/978-3-319-11505-4_5

**Table 5.1** Percentage of different land type among all converted farmland

| Year | Farmland with a slope greater than 25° | Desert land |
|---|---|---|
| 2005 | 35.11 | |
| 2006 | 39.72 | 16.68 |
| 2007 | 24.70 | 21.59 |

Source: SFA (2006–2008)

1. Farmland where soil erosion is severe (mainly farmland with a slope of more than 25°).
2. Farmland where desertification or alkalinity are severe.
3. Farmland located in an ecologically important area, where the capacity to farm productively is low and unstable.
4. Devastated land where soil erosion is severe.

Liu and Li (2010) argue that GfG regulation stipulates that cropland should not be converted if its productivity is sufficiently good to produce grain that commands a higher value than the GfG subsidies, and the land does not cause soil erosion. There is space for negotiation, however, if the conversion of such land is necessary for ecological reasons. As long as it is approved by the State Council, the planned area of conversion can be adjusted (Liu and Li 2010). Indeed, as discussed in Chap. 4, some scholars have pointed out that land has been misassigned, and therefore the cost-effectiveness has been overestimated. Table 5.1 provides an overview of the percentage of land converted in the first two categories. It shows that only about one third of the land converted through the GfG was farmland with slope higher than 25°, while the amount of desert land converted was even smaller.

Zuo et al. (2003) argued that what contributed to the misassignment of land was the fact that most village and township governments preferred the easier-to-implement method of simply targeting all steeply sloping cropland in the township rather than targeting areas based on the conditions of entire catchments. Officials generally took into account ecological conservation, watershed services and the types of vegetation appropriate for local conditions when delineating areas for program enrolment (Bennett 2008). There are also reports, however, that they tended to focus on retiring contiguous swaths of land to convert to forests because, by doing so, it was easier to implement and monitor the program; this might have included fertile flatland that was not the focus of the GfG.

The selection of the land to be set aside was done as follows. The Ministry of Forestry[3] issued targets for the total amount of land to be converted in each province, which in turn issued targets for each county, which issued targets for townships, which then issued targets for villages. For example, in 2002 the central government issued a target of 200,000 ha of land to be converted in Yunnan Province, half of which was to be reclaimed farmland and the other half forests planted on wasteland and barren mountains. The Yunnan provincial government issued targets for each

[3] In 1998 the Ministry of Forestry changed its name to State Forestry Administration (SFA) but in line with most of the literature on Chinese forestry we maintain the name Ministry of Forestry (MOF).

county. For example, Dayun County was to create 2,000 ha of forest, 1,000 of which were to be reclaimed farmland, and the other 1,000 were to be planted on wasteland and barren mountains. Forestry bureau staff from each county then went to villages and invited people to join (Delang and Wang 2013).

Each year there was a quota for each county (though not all counties were involved). In most cases, in particular in counties with much slope land, farmers working land with slopes steeper than 25° were invited first, but if there was not enough slope land to fill the quota, others were also invited to join. At the beginning, few people wanted to join because they did not trust the government. However, since payments were rather generous, made without fail, and the program was generally well organised, more households wanted to join. As a result, the quota was insufficient for all those who wanted to join (Delang and Wang 2013).

This chapter reviews the factors that determined selection of land, and considers the differences among provinces and smaller administrative units. In the next section, we look in more detail at the effectiveness of the program targeting in terms of steepness of the sloped land. In section "Land slope", we look at the suitability for farming of the converted land. In the last section, we consider the socio-economic conditions of the farmers and the extent to which the program targeted poor households first.

## Land Slope

The GfG program focused on western China because the area contains the headwaters of the Yangtze and Yellow rivers, and it accounts for around 80 % of the total area identified with soil erosion problems (>360 million ha). This region is critical not only to itself (for example, in relation to biodiversity and land degradation), but also to the middle and lower reaches of the two rivers, which cover almost half of China's territory (Feng et al. 2005b). According to the 1996 national land survey, a total of 38 million ha were cultivated in western China. This accounts for 28 % of all cultivated land in China. However, the West contains a higher percentage of cultivated slope land than the rest of the country (Feng et al. 2005b). Approximately 55 % of the land in this region has slope greater than 25° and is subject to severe soil erosion. This results in large amounts of silt released in the two rivers: in the early 2000s, two to four million tons of silt were released into the Yangtze and Yellow Rivers each year, 65 % of which was estimated to come from sloping cropland (Bennett and Xu 2005).

According to Xu and Cao (2002), 91 million mu (6.067 million ha) of cropland in China has a slope greater than 25°, more than 70 % of which are found in western China (Bennett 2008). Excluding Xinjiang and Tibet, slope land accounts for more than half of the total agricultural land in the western provinces. Reforesting these slope lands was an important policy to enhance the capacity for soil and water conservation. The intention was that, on completion of the GfG, 75 % of the sloping

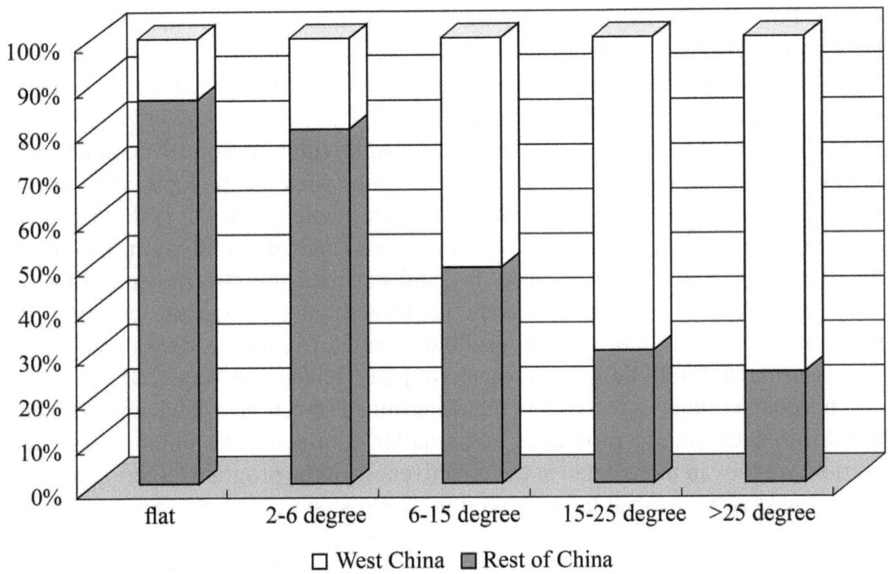


**Fig. 5.1** Share of cultivated sloping land in the West relative to all of China (Source: Feng et al. 2005b)

**Table 5.2** Area of cropland of different slope degrees in 100 sampled counties

| Year | 1998 | 2003 |
|---|---|---|
| Total cropland (mu) | 20.09 | 8.72 |
| Cropland with slope over 25° (mu) | 7.87 | 1.82 |
| Cropland with slope below 25° (mu) | 12.22 | 6.9 |

Source: SFA (2005d)

farmland along the upper stream of the Yangtze River and the Yellow River, and 46 % of farmland under desertification would be converted into forest land or grassland (Hori and Kojima 2008) (Fig. 5.1).

Even though slope gradient should be a criterion for the selection of agricultural land (or waste land), case studies have shown that practice is not always consistent with theory. For example, the SFA (2005d) reported that in 100 counties monitored by the government in 2003, there was a 77 % drop in the croplands with slopes greater than 25°. However, the amount of cropland with slopes below 25 % also decreased by 44 % (Table 5.2). The SFA (2005d) estimated that by 2004 croplands with a slope of 25° or above accounted for 40.06 % of the overall converted land. Hence, since the inception of the GfG, a number of studies have advocated better "targeting" land to be revegetated, in terms of land slope gradient (Uchida et al. 2005; Xu et al. 2010).

Land targeting in the GfG program has been studied particular by Uchida et al. (2005) and Xu and Cao (2002). In 2001, Xu and Cao (2002) looked at one county

**Table 5.3** Comparison of slopes from case studies in China's GfG Program in 2000

| Province | County | Percentage of land with slope ≥15° | |
|---|---|---|---|
| | | Cropland set aside | Cropland not set aside |
| Gansu | Dingxi | 83 | 45 |
| Inner Mongolia | Zuozi | 16 | 33 |
| Ningxia | Pengyang | 93 | 72 |
| Yunnan | Heqing | 96 | 91 |
| Guizhou | Dafang | 98 p | 69 |
| Sichuan | Tianquan | 86 | 65 |

Source: Adapted from Xu and Cao (2002)

in each of seven provinces (Gansu, Inner Mongolia, Ningxia, Shaanxi, Yunnan, Guizhou and Sichuan) to determine whether the land selected for GfG was actually slope land. In five out of the seven counties they reviewed, more than 80 % of the plots selected for Grain for Green had slopes of more than 15°. In some counties, the program had been very effective in focusing on slope land. For example, in Dafang County (Guizhou Province) 98 % of the plots enrolled in the GfG had slopes of more than 15° (Table 5.3).

Uchida et al. (2005) also looked at the steepness of converted land, drawing primarily on a dataset collected in 2000 that covered 144 participating households from 16 randomly selected villages in Ningxia and Guizhou Provinces. Data from a series of community surveys in six provinces were used to supplement the analysis. While they reached similar conclusions as Xu and Cao (2002), they also presented additional insights. Most notably, they concluded that while many of the plots selected for the program were steeply sloped, a fairly small share of the participating plots in the program were not sloped, and a fairly large number of steeply-sloped plots were not include in the program. For example, in Dingxi County (Gansu province) 83 % of the cropland set aside under the program was steeply sloped. Seventeen percent of the plots, however, were not.

Uchida et al. (2005) noted that, unfortunately, the survey did not ask the precise reason for plots to be excluded from the program. Therefore, they were unable to assess whether these plots were excluded because officials did not allow their inclusion or because farmers did not want to have them in the program. Uchida et al. (2005) tried to identify the reason for non-inclusion by looking at different options, such as plot characteristics, household characteristics, village characteristics and institutional factors. They concluded that, when households have greater decision-making power of implementation, they favor retiring plots with lower opportunity costs, which they may measure in different ways: lower land quality, greater distance to home, poorer irrigation conditions, more uncertain land rights. On the whole, they suggested that increasing household autonomy regarding whether to participate could improve program cost effectiveness by improving the likelihood that the plots of least cost for households would be chosen. Nevertheless, Uchida et al. (2005) show that, although there are some targeting problems, to a remarkable degree program officials are setting aside cultivated land with significant slope.

Similarly, Xu et al. (2010) used another 2003 household survey to examine GfG's implementation and impact. They found that land targeting had been strongly influenced by program goals, but that mistargeting also occurred. Xu et al. (2010) estimated that 21 % of the retired land in their sample were low-sloping plots (less than 15°), and that, on average, 71 % of this land in each village could have been replaced with unenrolled, highly sloping land. This indicates that considerations other than plot slope must have been important in the enrolment choice in these villages. To increase the environmental benefit of the GfG, Xu et al. (2010) suggested that the relatively flat plots should be replaced with more steeply sloped land currently not in the program. Such a swap would be fairly easy to accomplish logistically as nearly half of the remaining non-program plots in the study site were highly sloped. Using a treatment effects approach to evaluate program impact, they also found evidence of lack of participant choice.

Mistargeting might have occurred for different reasons. In some cases plots closer to roads have been targeted to "showcase" the implementation of the program to higher-level authorities (Zuo et al. 2003; Xu and Cao 2001). In other cases, as Xu and Cao (2002) reported, in addition to land with high slopes, some regions have given priority to sites close to a road system in order to facilitate inspections and monitoring. For example, Xu and Cao (2002) reported that in southwest China more than 70 % of the farm households in the program were located along a road. Yet in other cases, the China Council for International Cooperation on Environment and Development (CCICED) reports that some regions required the plots to be contiguous to each other to minimize implementation costs, which resulted in the inclusion of cropland that covered relatively flat areas (CCICED 2002, in Uchida et al. 2005).

## Land Productivity and Suitability for Farming

During the post-pilot phase in particular, it was evident that in some regions a significant portion of high-quality, low-sloping land was enrolled under the program, while high-sloping low-quality land remained in cultivation (Xu et al. 2004). Overall, however, the program has been effective in targeting plots with slope land. Xu et al. (2004) examined both slope and productivity of the land, using a dataset of 358 households collected in 2003 in the three western provinces where the GfG was first initiated on a pilot basis in 1999 (Gansu, Shaanxi and Sichuan provinces). This made it possible to look at program implementation over a comparatively long period of time.[4] The data include both participants and non-participants, hence allowing for examination of overall targeting of plots in the selected areas. The land is categorized as (1) Low, Medium, or High productivity, based on 1999 (i.e. pre-GfG) net revenue per hectare, and (2) High, Medium and Low Slope land, defined as land that has a slope greater than 25°, between 15° and 25°, and less than 15°, respectively.

[4]The 358 households were selected via stratified random sampling of two counties in each province, three townships in each county, two villages in each township and 10 households in each village. The dataset was collected by the Center for Chinese Agricultural Policy, Chinese Academy of Sciences.

**Table 5.4** Comparison of slopes from case studies in China's GfG Program in 2003

| | Percent of slope | | |
|---|---|---|---|
| Province | <15° | ≥15° and ≤25° | ≥25° |
| Gansu | 38 | 14 | 48 |
| Shaanxi | 10 | 27 | 63 |
| Sichuan | 11 | 14 | 75 |

Source: Xu et al. (2004)

The dataset does not account for labor inputs and household characteristics (Xu et al. 2004).

Xu et al. (2004) concluded that, while in general plots with higher slopes (Table 5.4) and lower productivity appeared to be predominantly targeted, significant variation existed. As much as 63 %, 48 % and 75 % of the converted plots in Shaanxi, Gansu and Sichuan, respectively, had slopes greater than 25° and were of medium to low productivity. This is encouraging. At the same time, however, in Gansu almost 19 % of the converted areas were low-slope and high-to-medium-productivity, while 38 % of converted land was in low-slope area. Furthermore, almost 10 % of the converted land in Shaanxi, and in Sichuan almost 11 % of converted parcels, were on land with a slope of less than 15°. This reveals problems in implementation since, in general, low-slope land should not be selected for the program, especially if it is relatively productive (Xu et al. 2004).

Another indication of good targeting is whether the share of the total area of lower-productivity and higher slope land that has been converted is greater, preferably much greater, than that of higher-productivity lower-slope land. For Shaanxi, almost 82 % of total high slope-low productivity land area has been converted, compared to less than 4 % of the high productivity-low slope land. However, in Sichuan, only 20 % of the high slope-low productivity land has been converted, whereas around 34 % of low slope-medium to high productivity land has been enrolled. In Gansu, only 56 % of high slope-low productivity land has been converted, while almost 10 % of low slope-high productivity land has been converted (Xu et al. 2004). In general, these results suggest that, at a minimum, far too much productive, low sloping cropland has been retired, particularly since in all samples high sloping, low to medium productivity land area remains in use (Xu et al. 2004).

Uchida et al. (2005) also described improvements in the selection of fields, once land productivity was included, and estimated potential savings in the compensation paid to farmers. Their findings were discussed in Chap. 3.

## *Suitability for Farming*

An issue that is closely related to whether the least productive land has been converted, is whether the land that has been targeted is also the least suitable for farming. To answer this question, Dong et al. (2010) tried to identify the spatial pattern of China's GfG by means of a land use change dataset using remote sensing, focusing

**Table 5.5** Valuation of the indicators of habitat suitability for cropping

| Aspects | Indicators | Suitability level | | | |
|---|---|---|---|---|---|
| | | High | Moderate | Marginal | Unsuitable |
| Climate | Annual precipitation (mm) | ≥600 | <600 | – | – |
| | AAT10°C | ≥4,000 | 3,400–4,000 | ≤3,400 | – |
| Topography | Slope (°) | ≤6 | 6–15 | 15–25 | ≥25 |
| Soil fertility | | ≥3.5 | 2.5–3.5 | 1.5–2.5 | ≤1.5 |

Source: Dong et al. (2010)

on what they considered to be a "typical" region: the middle and northern Shaanxi province in the Loess Plateau. The Loess Plateau, the cradle of Chinese civilization, suffers from the world's most intensive and extensive erosion. In recent decades economic development and land over-use have accelerated soil erosion due to the inherent fragility of loess soil. The GfG was implemented in Shaanxi Province in 1999. The middle and northern areas of Shaanxi province were the most "typical" regions in the Loess Plateau and were therefore chosen as the study area.

Dong et al. (2010) tested the spatial rationality of the selection of fields by looking at the suitability of habitats for cropping. To do so, they identified changes of crop-land between 2000 and 2005, comparing land use images of 1999/2000 to land use images of 2004/2005. Subsequent field investigation confirmed that the estimation accuracy was higher than 95 % (Liu et al. 2009). Dong et al. (2010) predicted the habitat suitability levels for cropping in a specific area through a conceptual model that related each measurable environmental variable to the degree of suitability of cropping. Four grades of habitat suitability were defined: high suitability, moderate suitability, marginal suitability and unsuitability. Plant growth was considered to be influenced by three external physical factors: climate, soil conditions, and topography (socio-economic factors were not considered) (Table 5.5).

1. Climate: This included annual accumulated temperature above 10 °C (AAT10) and precipitation, which was valued according to the demand of crops.
2. Soil condition: Soil fertility including organic matter, total nitrogen, total phosphorus, and total potassium. Dong et al. (2010) derived soil data sources from the Second National Soil Survey. The quality of these soil indicators was evaluated and expressed as scores, and comprehensive scores were calculated for each unit as weighted summations.
3. Topography: Slope gradient was considered to be the key limiting factor in GfG projects, and slopes above 25° were considered unsuitable for agriculture.

Subsequently, Dong et al. (2010) overlaid GfG maps from remote sensing to habitat suitability maps to find how much land was (a) unsuitable, (b) of marginal suitability, (c) of moderate suitability, and d) of high suitability for farming. Table 5.6 presents the results. Of the total area analyzed by Dong et al. (2010), 80.61 % was either marginally suitable or unsuitable for farming. Although only a small amount of land was unsuitable (3.26 %) the GfG seems to have been targeted fairly

**Table 5.6** Area distributions of GfG with different habitat suitability levels ($km^2$)

| | Converted land type | | | |
|---|---|---|---|---|
| Suitability level | Grassland | Forest | Total | Area percentage in total GfG area |
| High | 0.20 | 0.0 | 0.20 | 0.01 |
| Moderate | 162.03 | 121.70 | 283.73 | 19.38 |
| Marginal | 528.94 | 603.32 | 1132.25 | 77.35 |
| Unsuitable | 20.31 | 27.36 | 47.67 | 3.26 |
| Total area | 711.48 | 752.38 | 1463.86 | 100 |

Source: Dong et al. (2010)

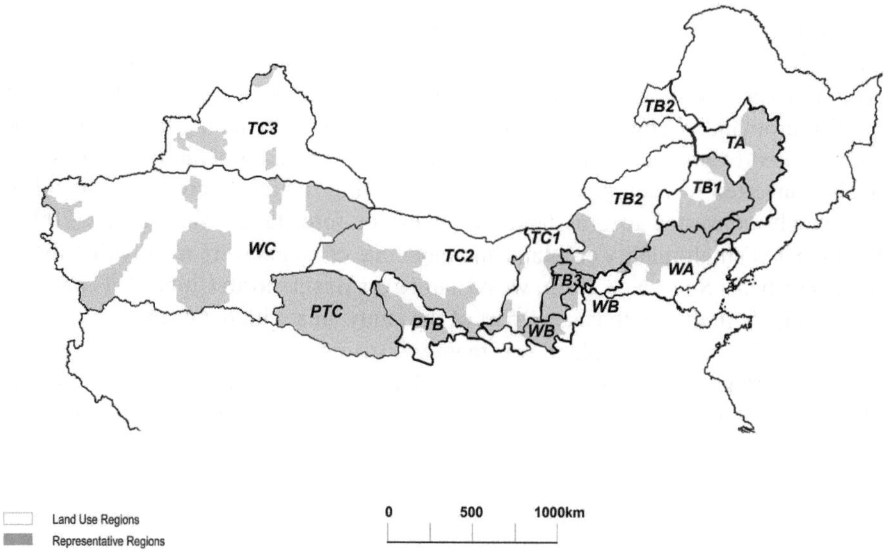


**Fig. 5.2** Division of land use in the desertification-affected north China (Note: TA = Northeast Song-Nen Plain; TB1 = West Liao River Basin; TB2 = Central Inner Mongolia Plateau; TB3 = East Ordos Plateau; TC1 = Ningxia Hetao; TC2 = Hexi and Alashan; TC3 = Junggar Basin; WA = North China Hills; WB = Northwest Loess Plateau; WC = Tarim Basin; PTB = East Qinghai Hills; PTC = Qaidam Basin. Source: Wang et al. 2007c)

efficiently, since 77.36 % was of marginal suitability only (Table 5.5). Approximately 19 % of the land that was considered of moderate suitability had been set aside, which is a relatively high number. However, only 0.01 % of the land converted by the GfG was highly suitable for farming. Dong et al. (2010) concluded that, on the whole, land selection had been rational, even though there was still room for improvement.

Wang et al. (2007c) also looked at the issue of the selection of land to be set aside, using a 2005 remote sensing survey of 395 counties in the areas in north China most affected by desertification. To do so, they divided the study area into 12 land-use regions (Fig. 5.2) by using factor analysis in combination with systematic data on land use (especially cropland use) at the county level that had been gathered in the first nationwide land survey in 1996. Apart from slope gradient, other factors that they took into account to evaluate cropland suitability included temperature conditions, soil erosion intensity, irrigation accessibility, soil drainage, soil depth, soil texture, degree of paludification, depth of obstacle horizon, and soil salinity. Socio-economic factors were not included in the analysis because they were considered less important than the physical factors in a study area with such fragile environment and relatively backward economy (Wang et al. 2007c).

Wang et al. (2007c) evaluated cropland suitability in selected representative regions that included 208 counties, covering an area of about 1.4 million km$^2$, occupying 35.5 % of the study area. Cropland suitability of a mapping unit was rated and attributed based on the principle of Liebig's law of the minimum with the help of GIS software. Cropland suitability was divided into four classes: "high quality cropland" (S1), "moderate cropland" (S2), "marginal cropland" (S3) and unsuitable cropland (N). Both moderate and marginal croplands were regarded as low quality cropland. Wang et al. (2007c) contended that high quality cropland should be protected to produce food for local people; moderate cropland could be selectively converted to ecological land use types, while marginal cropland should be completely converted to grassland, scrubland or forestland. Wang et al. (2007c) found that (1) high-quality cropland and moderate cropland were mainly distributed in the northeast Song-Nen Plain, west Liao River Basin, north China Hills and east Qinghai Hills in the south eastern part of the study area with better moisture conditions; and (2) marginal cropland was mainly distributed in the north western part of the Loess Plateau and around the central part of the Inner Mongolia Plateau (Table 5.7). Wang et al. (2007c) argued that all types of low-quality cropland, not

**Table 5.7** Cropland suitability by representative region (percent)

| Region | High quality cropland (S1) | Moderate cropland (S2) | Marginal cropland (S3) | Unsuitable cropland (N) |
|---|---|---|---|---|
| Northeast Song Neng Plain | 44.09 | 18.52 | 11.39 | 26.00 |
| West Liao River Basin | 17.85 | 12.35 | 13.00 | 56.80 |
| Central Inner Mongolia Plateau | 2.66 | 11.95 | 21.91 | 63.49 |
| East Ordos Plateau | 0.39 | 3.55 | 3.29 | 92.77 |
| Ningxia Hetao | 11.78 | 9.90 | 5.50 | 72.82 |
| Hexi and Alashan | 6.16 | 4.41 | 1.35 | 88.08 |
| Junggar Basin | 7.10 | 0.54 | 1.36 | 91.00 |
| North China Hills | 19.89 | 8.27 | 6.72 | 65.12 |
| Northwest Loess Plateau | 4.38 | 5.49 | 33.87 | 56.26 |
| Tarim Basin | 0.78 | 0.57 | 0.53 | 98.12 |
| East Qinghai Hills | 23.68 | 18.16 | 5.72 | 52.44 |
| Qaidam Basin | 0.21 | 1.23 | 3.01 | 95.55 |
| Total | 7.99 | 5.14 | 5.48 | 81.40 |

Source: Wang et al. (2007c)

only steep cropland, should be converted for ecological purposes. Furthermore, they called for a more complete conversion of marginal cropland because continued use of this cropland would increase desertification. The conversion of moderate cropland, Wang et al. (2007c) continued, should be implemented only gradually to avoid possible food insecurity problems triggered by quick conversion. Therefore, they specifically suggested that:

1. Both marginal and moderate croplands should be converted in the West Liao River Basin, Ningxia Hetao, Hexi and Alsshan, Junggar Basin and East Qinghai Hills, where high-quality cropland is sufficient.
2. All the marginal cropland and part of moderate cropland should be converted in the northeast Song-Nen Plain, central Inner Mongolia Plateau and Qaidam Basin, where high-quality cropland is limited and thus some moderate cropland is needed for meeting the food demands of local people.
3. Only marginal cropland should be retuned in the east Ordos Plateau, North China Hills, northwest Loess Plateau and Tarim Basin, where both high-quality cropland and moderate cropland are needed for food security and sustainable agricultural production.

Ultimately, all marginal and moderate croplands should be converted to grassland, scrubland or forestland in areas where food security is basically guaranteed (Wang et al. 2007c).

## Socio-economic Considerations

In addition to the possible mistargeting of its physical, environmental and ecological objectives, the cost-effectiveness of GfG from a socio-economic point of view has also been called into question from the beginning of its implementation. Gauvin et al. (2010) used survey data to evaluate what factors determined selection of program areas for the GfG program. To do so, they built a model that considered the dual goals (environmental restoration and poverty alleviation) of the GfG, and estimated the extent to which environmental and poverty alleviation goals could be achieved, given fixed subsidies. The authors used the calculated benefits and costs to evaluate the effectiveness of GfG on different parcels of land.

For simplicity, Gauvin et al. (2010) assumed that the program paid each household its opportunity cost as program compensation. How much of the environmental and poverty alleviation goals can be achieved simultaneously, given a fixed budget, depends on the degree of trade-off between the gain of the two benefits obtained when retiring a parcel. If environmental and poverty alleviation benefits involve tradeoffs, the program manager must decide how much weight to assign to environmental benefits relative to poverty alleviation benefits. The implication of this key trade-off is that utilizing a targeting approach that is suitable for reaching a program's environmental goals does not necessarily allow the program to simultaneously reach its poverty alleviation goals. In contrast, if there is not a sharp trade-off between the two goals, then retiring a parcel which has high environmental benefits would also further the program's poverty alleviation goals (Gauvin et al. 2010).

Gauvin et al. (2010) used the following methodology rationale. First, they econometrically examined what factors affected land selection in their study areas. Based on the econometric findings, they tested whether or not the environmental benefits, the opportunity cost and the poverty levels of the households were considered. Next, they used land data to graphically examine heterogeneity across their study areas in terms of the environmental benefits, the opportunity cost and the poverty levels of the households in the sample. Gauvin et al. (2010) also examined the correlations among these three factors.

The study combined two datasets. The first dataset was collected through household surveys that they designed (commissioned by China's Ministry of Forestry) and administered in 2003, approximately 3 years after the start of the program. The second data source included interviews with a total of 359 households in 36 villages.[5] The survey asked respondents for information regarding their situation prior to entering the program (1999) as well as after the program (2002). Information was collected at both the household and land levels, including detailed information on each household's total asset holdings, its demographic makeup, net revenue per mu, and other income earning activities (Gauvin et al. 2010).[6]

Based on the descriptive statistics, both the opportunity cost of retiring the parcel and the environmental benefit of land retirement appear to have been factored into targeting. After plotting the asset levels of the households against plot-specific opportunity costs, Gauvin et al. (2010) concluded that selection could have been better in terms of targeting lands belonging to poor households with lower opportunity costs (Fig. 5.3a). In Fig. 5.3a, data points of GfG participants gather at the left-bottom of the plot, showing that most GfG participants had low asset levels and the

[5] Distributed in three provinces (Sichuan, Shaanxi, and Gansu), six counties (two for each province), and 18 townships (three per county). The 36 villages accounted for two villages for every township. Ten households within each village were randomly selected. At least one participating household was selected in every village. In two out of the 36 villages, all of the selected households were participants of the program (Gauvin et al. 2010).

[6] Gauvin et al. (2010) used the level of assets prior to the start of the program to measure each household's pre-program wealth, instead of using income as the indicator since the latter is often subject to substantial measurement error. A household's total assets were calculated as the sum of the value of each family's house in 1999, the total value of 18 consumer durables and the total value of 18 fixed productive assets. The opportunity cost of participating in the program was calculated as the net revenue of the land one year prior to entering the program. The net revenue per mu was equal to the gross revenue per mu minus the plot's variable costs per mu, which included expenditure for fertilizer, pesticide, plastic sheeting and hired labor. The cost did not account for the value of household labor. They also investigated land area (mu), the distance to the house (km) and an index of slope. Their measure of the environmental benefits of retiring cultivated land attempted to capture the potential reduction in soil erosion by combining information from their survey on land-specific slope and water erodibility index from a national database.

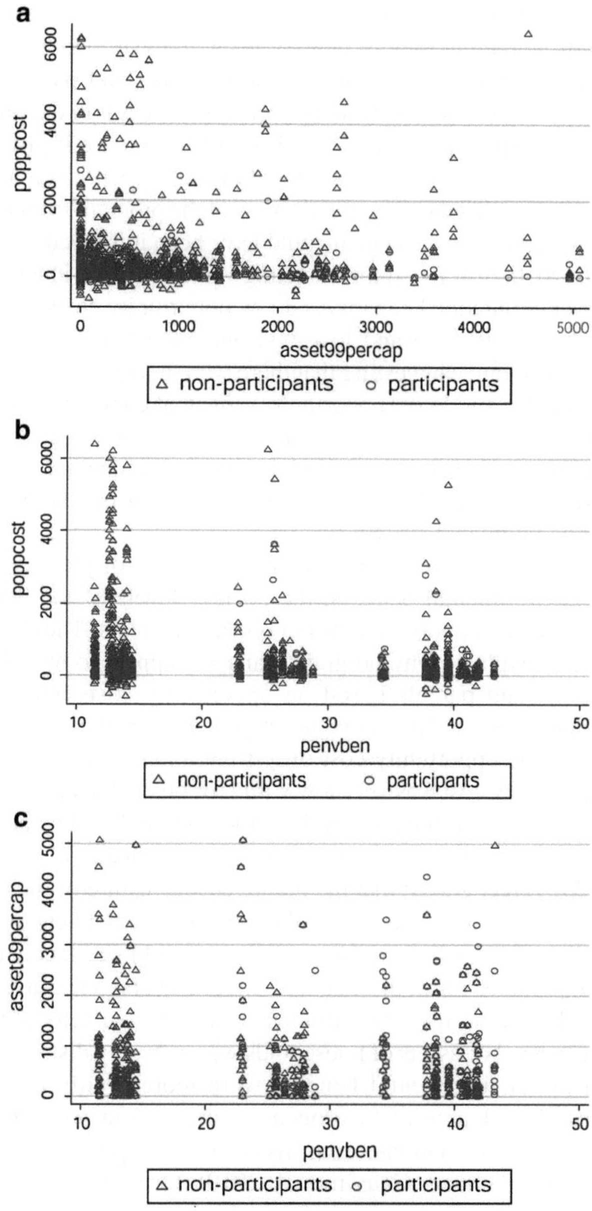


**Fig. 5.3** Scatter plots: (**a**) asset level per capita versus opportunity cost of each plot of land that is enrolled and not enrolled in the program, 2002; (**b**) land environmental benefit versus opportunity cost of each land that is enrolled and not enrolled in the program, 2002; (**c**) asset level per capita versus environmental benefit of each land that is enrolled and not enrolled in the program. *Triangles* represent the non-participants' plots and *circles* represent the program participants' plots (Note: poppcost: opportunity cost per unit area; penvben: environmental benefit; asset99percap: household asset per capita in 1999. Source: Gauvin et al. 2010)

opportunity cost of their land was low. Thus, GfG selection was effective in terms of targeting lands belonging to poor households with lower opportunity costs. The researchers also found, however, that there were many land parcels managed by poor households that had low opportunity costs but were not in the program (lower left hand quadrant). That land would have been ideal to retire if program mangers wanted to maximize the program's poverty alleviation goal, given a fixed budget. They argued, therefore, that the cost effectiveness of the program from the perspective of reaching its poverty alleviation goal could have been improved.

Gauvin et al. (2010) also found that targeting was not perfect in terms of targeting lands with high environmental benefits and low opportunity costs (Fig. 5.3b). They found that lands of both high and low environmental benefits were selected to participate in the program, suggesting that there were numerous parcels with higher environmental benefits and lower opportunity costs that could have been enrolled in the program, but were not selected. In addition, Fig. 5.3c suggests that program managers could have selected parcels with higher environmental benefit that were managed by poorer households, which would have better met both program goals simultaneously (Gauvin et al. 2010).

Gauvin et al. (2010) also examined the heterogeneity of plots in the sample in terms of their environmental benefits, the poverty levels of the households that control the plots and the opportunity cost of retiring the plots. The degree of heterogeneity[7] ultimately explains why program managers appear to have been selecting areas, households, and parcels based on opportunity costs and environmental benefits, but not poverty. The most heterogeneous of the three issues under consideration was the opportunity cost of retiring a land parcel (that is, for some parcels, the opportunity cost was very low while for others the opportunity cost was very high), followed by the poverty levels of the households. The parcels were the least heterogeneous in terms of their environmental benefit (Fig. 5.4). However, Gauvin et al. (2010) recognized that the homogeneous nature of the environmental benefits could be driven by two features of the environmental benefit index: (1) the national erodibility index was shared across all plots within each county; and (2) the plot-specific slope index was a discrete and thereby coarse measure. Furthermore, they found that there were important differences in heterogeneity across provinces (Fig. 5.5a–c). The level of assets of households was unevenly distributed in all three provinces and the environmental benefits were more evenly distributed in all three provinces. The degree of heterogeneity differed in terms of opportunity cost - Shaanxi had a more homogeneous distribution of opportunity cost, followed by Gansu and lastly Sichuan (Gauvin et al. 2010). Overall, these figures indicate that in order to select parcels more effectively the authorities should focus on

[7] The degree of heterogeneity is a statistical method that describes the differences within a set of data. Here, it refers to the difference in opportunity cost and environmental benefits among various parcels of converted land.

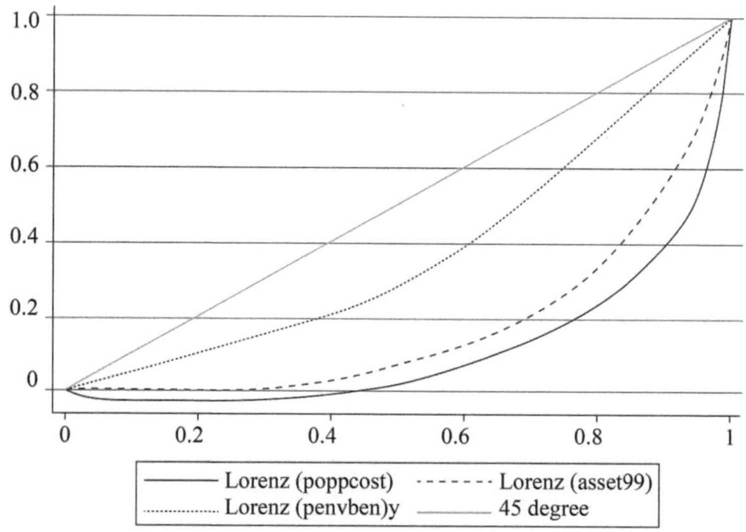


**Fig. 5.4** Proportion of parcels' opportunity cost, asset level, and environmental benefit achieved given proportion of parcels, 2002 (Note: The *curves* indicate the Lorenz curves for opportunity cost per unit area (poppcost); environmental benefit (penvben); and household asset per capita in 1999 (asset99). The *straight line* represents the 45° line. Source: Gauvin et al. 2010)

more carefully targeting the level of assets of households and the opportunity cost of retiring a plot.

In sum, Gauvin et al. (2010) found that there would have been a substantial gain in the cost effectiveness of the program if the land with the lowest opportunity cost (managed by poorer households) and highest environmental benefit had been targeted. However, they also pointed out that implementing this "golden rule" may be particularly challenging in developing countries because there are often poor data and low levels of institutional capacity. Since the environmental benefit of a plot appears to be the most homogeneous of the variables targeted, the program manager dealing with a fixed budget would gain more by making an effort to collect information on the other two targeted variables: the opportunity cost of taking the parcel out of production and the asset values of the owners of the plots (Gauvin et al. 2010). These results, as indicated by the authors, may be driven in part by the way they measure the three dimensions of the program, especially when the measure of environmental benefit (1) merely examines the reduction soil erosion, (2) is not directly measured in the field and (3) excludes other significant factors in assessing soil erosion.

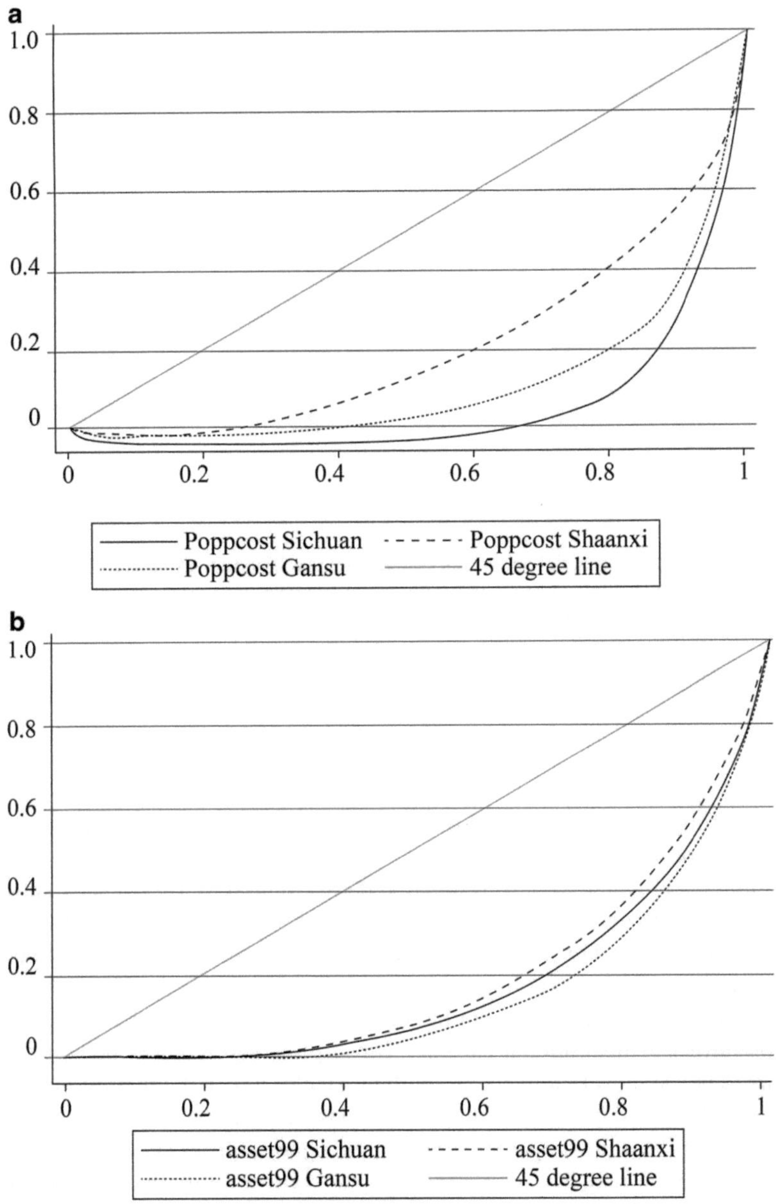


**Fig. 5.5** (**a**) Proportion of parcels' opportunity cost, achieved given proportion of parcels; (**b**) proportion of households' asset level, achieved given proportion of parcels; (**c**) proportion of parcels' environmental benefit, achieved given proportion of parcels (Note: The *curves* indicate the Lorenz curves for opportunity cost per unit area (poppcost); environmental benefit (penvben); and value of household asset per capita in 1999 (asset99). The *straight line* indicates a 45° line. Source: Gauvin et al. 2010)

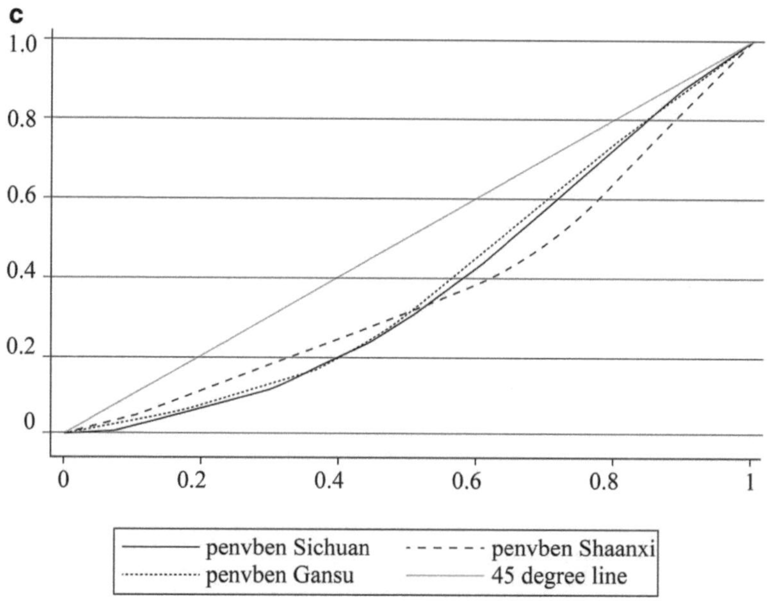


**Fig. 5.5** (continued)

## Conclusions

This chapter has reviewed considerations of land selection including how closely land selection adhered to program goals. The most important consideration in land selection was the slope level. However, slope was not the only factor. The fertility of the soil, its state of erosion or propensity to erode, and the incomes that it could generate for farmers were also important factors that program planners should have taken into consideration. The evidence reviewed in this chapter confirms that overall program targeting in terms of land selection has been fairly good, although in many places it could have been improved as a relatively large amount of non-slope land was included. In particular, the socio-economic characteristics (e.g. the poverty level) of the participants were not always fully taken into consideration, even though it is also clear that there was often insufficient information on the socio-economic characteristics of households to better target the program, and that obtaining such information is expensive.

# Chapter 6
# Plant Selection

**Abstract** This chapter discusses plant selection. The Grain for Green promotes the planting of either economic trees (trees from which a regular income may be obtained from the sale of non-timber products, such as fruits), ecological trees (trees that may be logged), or grassland. More farmers prefer to plant economic trees, because they generate higher and more regular incomes than ecological trees. However, the national standard is for ecological trees to make up 80 % of the total, and this is generally adhered to. In many places, farmers also claim that they do not have a choice of which plants to grow, but can only select from a few species.

**Keywords** Economic trees • Ecological trees • Tree species • Grass • Survival rate • Land ownership • Tree ownership

## Introduction

The Grain for Green program promotes the growth of three categories of plants: economic trees, ecological trees, and grassland. The distinction between these three plants is important because the three kind of plants command different levels of subsidies, for a different period of time. The plants also generate different levels of income for the farmers. First, we introduce the different kind of plants supported by the GfG, discuss their characteristics and the prevalence of different forest types. Second, we discuss the survival rate of different species, and the rules regarding survival rate and subsidy payments. The survival rate of the trees is officially very high, but many scholars have questioned the reliability of the official data. Third, we discuss the reforms that have taken place concerning land and tree ownership, and the importance of these reforms for the GfG.

## Plant Type

The GfG supports the regeneration of the original vegetation, which can be either grassland or trees. The trees planted can be either economic trees or ecological trees. Economic trees are those from which a sustainable income can be generated,



C.O. Delang, Z. Yuan, *China's Grain for Green Program*,
DOI 10.1007/978-3-319-11505-4_6

for example, through the sale of non-timber products such as tea, fruits, or nuts. These cannot be cut unless they stop producing marketable products, in which case they can be replaced with other economic trees (Li 2002). In contrast, the goal of ecological trees is to reduce soil erosion and sandstorms (Li 2001, 2003). Ecological trees (which include such species as fir, cedar, and other coniferous trees) may be logged, subject to quota and forest officers' approval (Delang and Wang 2012).

There are official standards for ecological and economic trees. In particular, the local forestry bureaus have lists of trees that may be planted with the financial support of the GfG, and usually each village has only a few species to choose among. The list of ecological trees includes 142 species in the southern region and 103 species in the northern region, while the list of economic trees includes 41 species in the southern region and 28 species in the northern region. Some tree species (such as walnut, chestnut, and tea) are included in both categories. Therefore, it is sometimes not possible to tell from the tree species whether the area has been planted with ecological or economic trees. There are also strict planting standards, including the density with which plants can be planted, rules to avoid soil erosion (such as planting hedges), the use of mulch, the frequency of weeding, and the areas that may be reclaimed. When ecological and economic trees are planted, some original trees may be cut to improve the ability of the new trees to prevent soil erosion (MOYN 2009: 141–151).

When the government designed the program, it was stipulated that in every administrative unit 80 % of the trees should be ecological trees and 20 % should be economic trees. This limit exists because economic trees, compared to ecological trees, require more frequent replanting and provide (in some cases) fewer environmental services. More frequent replanting may compromise the primary objective of the program: reducing soil erosion (Uchida et al. 2005).

However, this rule was not always enforced, as farmers preferred to plant economic trees rather than ecological trees (Cui 2009; Bennett et al. 2011). Farmers prefer economic trees because they can earn higher incomes from their fruits and other non-timber products, then use or sell the wood (from thinning) and timber, which may be harvested when fruit trees stop producing. Wang and Maclaren (2011) directly addressed farmers' preferences for particular tree types, through a survey in Dunhua County (Jilin province). They found that the design of the GfG did not reflect the needs and attitudes of the residents, with 66 % of survey respondents stating that their priorities differed from those of the government. Given the choice, 40 % of the farmers would have preferred to plant economic trees; 26 % of the farmers would have opted for "timber trees" (by which they presumably mean ecological trees) and 31 % favored ecological trees, bringing the total number of farmers who wished to grow ecological trees to 57 % – well below the government target of 80 %.

Similar findings were reported by Uchida et al. (2005), who evaluated the future profits of GfG plants, using a survey conducted in 2,000 among 144 participating households from 16 randomly selected villages in two provinces, Ningxia and Guizhou. While the actual implementation in Guizhou was consistent with the government's requirement, the survey shows that more than 50 % of households

**Table 6.1** Percentage of ecological forest area converted from cropland

| Year | Ecological forest (%) |
|---|---|
| 2002[a] | 93.42 |
| 2003[b] | 79.91 |
| 2004[b] | 80.84 |
| 2005[b] | 83.64 |
| 2006[b] | 79.64 |
| 2007[b] | 86.22 |

Source: [a]Data from 50 sampled counties (SFA 2003d); [b]National data (SFA 2003–2008)

stated that they would have preferred to plant economic trees (Uchida et al. 2005). Uchida et al. (2005) argued that if farmers had been allowed to plant economic trees, not only would they have had greater incentives to manage the trees more intensively, but they would also have been able to create an alternative income source that could reduce the propensity for reconversion when the subsidies end. Because of the high proportion of relatively economically, non-productive ecological trees, there may be a greater danger of reconversion in the future when program payments cease. Indeed, according to Li (2009b) the question remains as to whether in the long run, most species will generate sufficient economic returns, so that the removal of the subsidies will not alter the future financial situation of the farmers.

According to Ke (2007), if economic forests exceed 20 % in the targeted area, there would be no grain and cash subsidy paid for the additional trees. This may have contributed to government regulations being followed. Between 1999 and 2003, the program converted 914,500 ha of cropland and afforested 925,000 ha of land. Of the converted lands, 85.29 % were converted to ecological trees (Trac et al. 2007). Table 6.1 also shows that, from 2003 to 2007, ecological trees were often planted more frequently that the government had stipulated, although some species are classified as both economic and ecological trees, which makes reading such statistics more difficult.

## *Low Diversity of Tree Species*

Vegetative cover and forested area have increased considerably thanks to the GfG. However, the ecological efficiency of the GfG is often criticised on two grounds:

1. The species planted are very often not natives to the areas in which they are planted.
2. The choice of tree species depends to a large extent on the climate and soil conditions, which vary greatly in mountainous area. The specific species planted do vary among different townships, but there is very little diversity of tree species planted in each particular township, with excessive emphasis on a very small number of species, effectively creating monocultures.

**Table 6.2** Tree plantation areas of sampled peasant households in Liping

| Tree species | Area (ha) |
|---|---|
| Chinese fir | 66.68 |
| Masson pine | 51.37 |
| Bamboo | 31.31 |
| Pear | 8.63 |
| Tea | 7.77 |
| Orange | 5.19 |
| Oil teaseed | 5.91 |
| Tuliptree and hackberry | 4.69 |
| Sawtooth oak | 3.68 |
| Chestnut | 1.78 |
| Wild pepper | 0.82 |

Source: Zhou et al. (2007)

The literature abounds with examples of limited diversity in the number of species introduced by the GfG. For instance, in Dunhua county in southeastern Jilin Province, 96 % of the trees were Olga Bay larch (*Larix Olgensis Henry*). In Jiangxi Province, 60 % of the converted land in 2006 was planted with Oil Camellia. In Henan Province during 2000–2005, poplar accounted for 40 % of the reforested area, whereas other ecological tree species accounted for less than 2 %, and fruit trees were planted on the remaining area (Liu et al. 2008). Similarly, Zhou et al. (2007) surveyed the tree species planted in Liping County (Guizhou province), and found concentrations of a few species, with Chinese fir planted on 36 % of the land, Masson pine on 27 %, and Bamboo on 17 %. Another eight species shared the remaining 20 % of the reforested area (Table 6.2). Hong and Li (2000) surveyed the vegetation introduced through the GfG in Yulin County (Shaanxi province). They identified five species in Y village, three of which were ecological trees (Chinese arborvitae (*Platyclaudus orientalis*), Pea tree (*Caragana psammophyla*) and Chinese pine (*Pinus tabulaeformis*), and two of which were economic trees (Chinese jujube (*Zizyphus jujuba*) and Chinese apple (*Malus pumila*)).

GfG-supported reforestation is less diverse than the local forest, or even abandoned farmland. A survey in five randomly selected counties (Jingbian, Ansai, Baota, Yanchang, and Luochuan) in northern Shaanxi Province suggested that the number of plant species in plots where cultivation was abandoned was of 21–31 species, compared with a range of 9–14 species in the afforestation plots (Cao et al. 2009a).

The low species diversity of GfG planted forests may call into question the ecological success of the GfG. Forests may not be sufficiently diverse to support a diverse wildlife. Also, the monoculture is vulnerable to ecological disasters because of high exposure to the possibility of pests or fires (Wang and Maclaren 2011). On the other hand, one should acknowledge that low species diversity makes it easier for the farmers to take care of the trees. Low species diversity also helps the farmers generate higher economic returns due to economies of scale, which encourage them to look after the trees.

## *Choice of Vegetation: Trees Versus Grass*

Another criticism made about the GfG is that, in contrast to its enthusiasm for planting trees, the MOF, which is in charge of implementing the program on behalf of the central government, has shown less interest in other measures, such as grassland recovery, terracing, the construction of check dams, or other engineering measures, even if they are better suited in certain environments (Yin et al. 2013). As stated by the Forest and Grassland Taskforce of China (2003, p. 3):

> Implementation regulation has not been tailored to local conditions, and there has been an overemphasis on tree planting rather than restoring original vegetation cover. The SLCP does not give sufficient consideration to the ecological and economic functions of grasslands in semi-arid areas and the need to restore these ecosystems.

Populus is a species commonly used by the GfG, but it has been singled out as inadequate, specially in arid or semiarid areas. Observers have voiced their concern that planting poplars as a major species for forestation in arid and semiarid regions is problematic, given the limited precipitation. Populus is a fast-growing species with low water-use efficiency. It is hard to establish the trees in many conditions and wherever they are established, their deep root system can haemorrhage ground water through transpiration, lowering the water table and making it harder for native grass and shrubs to survive (Normile 2007). Many studies have reported that when the consumption of rainwater by tree plantations is higher than the level of consumption by natural vegetation, increased forest cover reduces the net runoff from a watershed (Cao et al. 2007a). Research in northern China (Wang et al. 2003) revealed that the runoff from afforestation plots decreased by an average of 77 % (ranging from 57 to 96 %) compared with grassland and farmland. Although this decreased runoff suggests increased retention of precipitation and decreased water erosion, the retained moisture is often used more rapidly than it can be replenished during the rainy season. As a result, the trees actually decreased the below ground water supply and the supply of water to rivers. Further, any soil conservation achieved by the trees was subsequently offset by more severe wind erosion (Cao 2008). Since 1949, the overall survival rate of trees planted during afforestation projects has been only 15 % across arid and semiarid northern China (Cao 2008).

## Survival Rate of Plants

Official figures commonly place, survival rates above 90 %, and even reach 100 %. These data are unrealistic when compared to normal survival rates from plantations. There may be two reasons for such a high percentage: First, local Forest Bureau officers may falsify the data to improve their performance. Second, farmers can replant every tree that has died and receive compensation retroactively. Compensation is conditional on the growth of the forest. Officers from the Forest Bureau verify the survival rate of the trees and the farmers must achieve a survival rate of 70 % (in the Yellow River watershed) to 85 % (in the Yangtze River watershed), now revised to a nation-wide standard of 75 %, to receive compensation (Bennett 2008). Farmers who do not achieve a survival rate of 70–85 % are allowed to replant the seedlings, and if the seedlings have survived when the officers from the Forestry Bureau inspect the fields the following year, the farmers are paid retroactively (for the previous year and the present one) (State Council 2007). The fact that the farmer can replant every tree that has died, and receive compensation retroactively, pushes the survival rate much higher.

Studies have shown that, in reality, survival rates are often below the government-stipulated level. Li (2009c) found that the survival rates of trees in many of the townships he surveyed were well below the standard stipulated by the government. Similarly, Bennett (2008) argued that the survival rates of planted trees in many of the townships in his dataset fell below those stipulated for subsidy delivery, although the low survival rates generally did not result in withholding subsidies. As Zuo et al. (2003) and others have observed during the pilot phase, program managers were faced with a dilemma when deciding whether to withhold subsidies, because of the program's dual goals of environmental restoration and poverty reduction. On the one hand, withholding subsidies based on low survival rates could dampen enthusiasm in the program, and reduce the number of people willing to participate. It would also harm the welfare-enhancing objective of the program. On the other hand, delivering the subsidies without adhering to the standards of compliance would encourage poor implementation by the farmers. Indeed, the failure to enforce the rule regarding no payment for low survival rates, could result in a vicious circle, which would lead to gradually lower survival rates.

To examine the determinants of survival rates of program-planted trees and grasses, Bennett et al. (2011) used a 2003 survey that collected household and plot-level data during and just after the pilot-phase of the program in the three initial pilot provinces: Shaanxi, Gansu, and Sichuan.[1] This dataset is used to examine the factors affecting the survival rates of program planted trees and grasses at the time of the first inspection. Figure 6.1 and Table 6.3 present the sample distribution of survival rates and tree/grass types. Survival rates in the Yellow River watershed area sample were mostly above the level stipulated by the government for the provision of subsidies (70 % of planted trees and grasses in the Yellow River watershed or north China), but in the Yangtze River watershed area it was often below the rate stipulated by the government (85 % survival rate in the Yangtze River watershed and south China) (SFA 2001a). As mentioned, more recently the survival rate required to obtain subsidies has been revised to a nationwide standard of 75 % (Bennett 2008).

Table 6.4 presents descriptive statistics regarding survival rates, tree and grass types planted, and enrolled area. Crops planted on the plots are grouped according to the Ministry of Forestry's program categories of "grasses", "economic forests" (orchard crops or trees with medicinal value) and "ecological forests" (timber crops)

[1] In 1999 and 2002, the survey collected detailed data for 360 households (including GfG participants and non-participants) in 36 participating villages on various household characteristics, off-farm income sources, plot-level agricultural inputs and outputs, husbandry and sideline activities, fixed and productive assets, and savings and credit. In total, 455 enrolled plots (of 246 participant households) were inspected at least once, and survival rate data were recorded (Bennett et al. 2011).

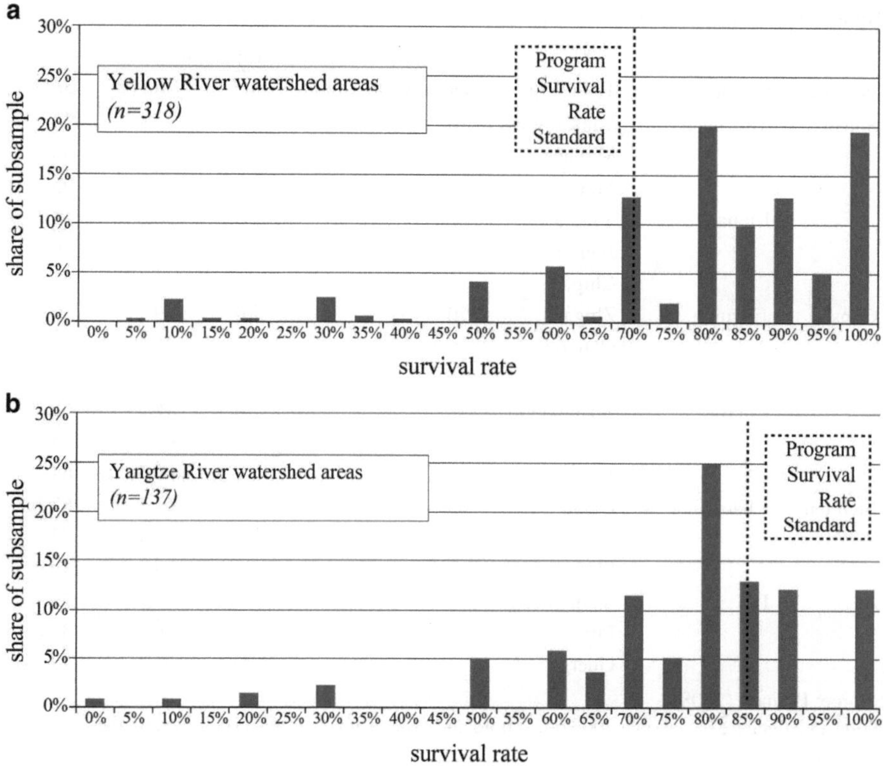


**Fig. 6.1** Histogram of sample survival rates, first inspection (Source: Bennett et al. 2011)

(Bennett et al. 2011). As Table 6.4 suggests, mean survival rates are not statistically different between "ecological forests" and "economic forests", while survival rates for trees are lower than those for grasses, significant at 1 % (Bennett et al. 2011).

In order to further examine the dynamics of tree survival, Bennett et al. (2011) developed a Tobit model with a number of different variables. Survival rates on first plots initially declined as the first trees succumbed. However, survival rates improve roughly 1 year into the program, as might be expected under a learning-by-doing scenario with replanting. Moreover, with the exception of the first 3 months of the program, survival rates on first plots are much lower than those of subsequently retired plots (Fig. 6.2) (Bennett et al. 2011). These results imply that it is important for each household to enter the program gradually, rather than being asked to retire a large portion of their land at the outset. They also suggest that agricultural extension programs need to accelerate the learning process, so that program benefits are delivered sooner (Bennett et al. 2011).

Bennett et al. (2011) found that households with higher cropping and husbandry income per capita also have higher tree survival rates. This suggests that households with higher agricultural labor productivity also perform better in forestry and horticulture.

**Table 6.3** Average survival rate of trees planted under the GfG (percentage)

| Province | County | Township | Govt. standard (%) | Inspections 1st (%) | 2nd (%) | 3rd (%) |
|---|---|---|---|---|---|---|
| Shaanxi | Yanchuan | Yanshuiguan | 70 | 94.2 | 93.6 | 98 |
| | | Majiahe | | 72.9 | 95.8 | 96.4 |
| | | Yuju | | 79 | 83.2 | 95 |
| | Liquan | Yanxia | | 56.3 | 86.8 | 81.1 |
| | | Jianling | | 78.8 | 47.9 | 39.4 |
| | | Chigan | | 100 | 46.7 | 52.1 |
| Gansu | Jingning | Zhigan | 70 | 70 | 69 | 66 |
| | | Gangou | | 80 | 76.6 | 71 |
| | | Lingzhi | | – | 75.7 | 77.7 |
| | Linxia | Zhangzigou | | 56.3 | 46.7 | 65 |
| | | Tiezhai | | 90 | 61.1 | 75.8 |
| | | Hexi | | 87.5 | 69.5 | 64 |
| Sichuan | Chaotian area | Datan | 85 | 82 | 61.5 | 67.3 |
| | | Zhongzi | | 70 | 48.7 | 77 |
| | | Shahe | | 92.5 | 74.1 | 40.4 |
| | Li | Shangmeng | | 100 | 79.6 | 76.1 |
| | | Puxi | | 74.9 | 80.7 | 84.8 |
| | | Guergo | | 70 | 74.1 | 77 |

Source: Bennett (2008)
Note: The data are based on a 2003 household and village-level survey conducted by the Center for Chinese Agricultural Policy

**Table 6.4** First inspection survival rates of program-planted trees and grasses

| Tree types | Survival rate, first inspection | | Share of enrolled area in sample (Total 89.7 ha) |
|---|---|---|---|
| | Mean | Std (%) | (%) |
| **Grasses** | 88.1 | 18.7 | 11.4 |
| Alfalfa | 93.5 | 9.5 | 5.3 |
| Ryegrass | 100 | | 0.1 |
| Chinese Toon (an herb) | 65.0 | 25.6 | 1.8 |
| Other grasses | 96.3 | 10.6 | 4.4 |
| **Economic forests** | 75.9 | 21.7 | 63.6 |
| Apple | 76.7 | 11.5 | 1.0 |
| Pear | 61.0 | 35.5 | 1.1 |
| Almond | 77.9 | 19.7 | 19.3 |
| Peach | 72.2 | 17.4 | 6.6 |
| Jujube | 75.8 | 28.7 | 20.7 |
| Prickley ash | 76.7 | 19.7 | 7.3 |
| Ginko | 86.7 | 15.3 | 1.2 |

(continued)

**Table 6.4** (continued)

| Tree types | Survival rate, first inspection | | Share of enrolled area in sample (Total 89.7 ha) |
|---|---|---|---|
| | Mean | Std (%) | (%) |
| Sumac | 65.0 | 7.1 | 0.2 |
| Mulberry | 78.8 | 18.9 | 1.4 |
| Sandthorn/Sea Buckthorn | 75.2 | 13.3 | 3.6 |
| Guava | 50.0 | | 0.1 |
| Persimmons | 30.0 | | 0.2 |
| Plum | 91.0 | 5.7 | 0.5 |
| Chinese arborvitae | 80.0 | 0 | 0.3 |
| **Ecological forests** | 75.7 | 22.4 | 25.0 |
| Black locust | 77.8 | 17.7 | 14.1 |
| Cypress | 83.1 | 17.9 | 2.6 |
| Willow | 87.2 | 8.8 | 1.5 |
| Japanese blue oak | 100.0 | | 0.0 |
| White poplar | 61.3 | 32.7 | 1.5 |
| Fir | 52.0 | 43.1 | 0.7 |
| Spruce | 85.0 | | 0.1 |
| Horsetail pine | 60.8 | 30.9 | 1.5 |
| Chinese ash | 50.0 | | 0.1 |
| Japanese black pine | 78.0 | 11.4 | 1.4 |
| Other tree types | 84.0 | 19.2 | 1.3 |

Source: Bennett et al. (2011)

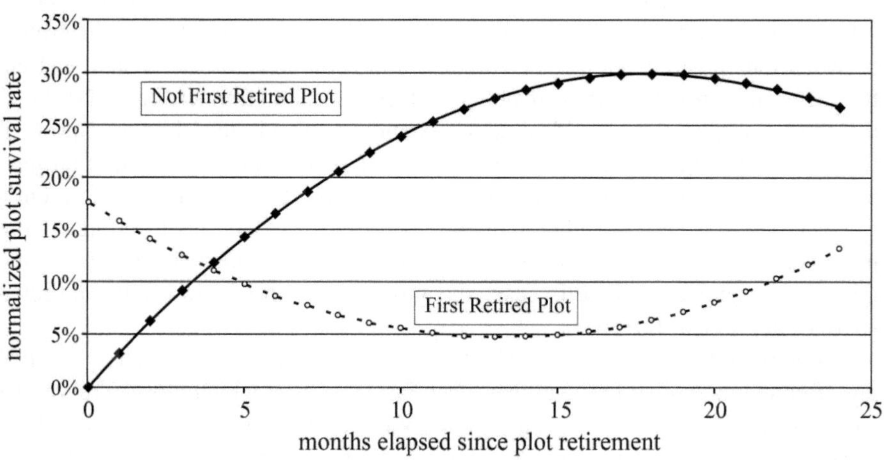


**Fig. 6.2** Estimated dynamics of survival rates (Note: Normalized survival rates are estimated from the Tobit model and are defined to be zero at the time of retirement for plots belonging to experienced households. Source: Bennett et al. 2011)

Bennett et al. (2011) also found that, perhaps contrary to expectations, the survival rates of plants by farmers who join the program voluntarily are similar to those who were instructed to do so by village leaders. This result offers two potential interpretations. It might mean that voluntarism has no particular impact on tree survival. Or it might mean that households that said they could choose whether to participate in the program, do not have substantially different rights from those who said they did not have a choice. This would occur if leaders were hesitant to press unwilling households into joining the program, or if households were reluctant to make decisions that would displease leadership (Bennett et al. 2011).

On the other hand, Bennett et al. (2011) argued that farmers who are permitted to choose what types of trees to plant obtain significantly higher survival rates – around 9 % higher. This result is obtained despite controlling for the types of trees planted, so the higher survival is not due to farmers selecting hardier ones. Rather, it implies that farmers who choose to plant a particular tree species are better able than the village leader to select tree-types that are more likely to survive, given plot characteristics and household constraints. Also, it is very likely that when farmers have the autonomy to choose what to plant, they have an increased propensity to invest effort and money into sustaining the plantation (Bennett et al. 2011).

Table 6.5 provides evidence in support of these interpretations. It shows that households with the right to choose what species to plant generally invest considerably more cash and labor on the plots they retire. These results suggest that farmers who can choose what to plant are more invested in the success of the retired land. This in turn suggests that granting farmers the right to choose the species is likely to align their interests more closely with the environmental goals of the program (Bennett et al. 2011). Table 6.5 also shows that, given a choice, farmers opt to grow economic trees, rather than ecological trees. This tendency is statistically highly significant, and likely the result of the fact that households derive economic benefits from economic forests much sooner (Bennett et al. 2011). The result carries two possible interpretations. First, the difference between the subsidies paid for ecological and economic forests is smaller than the difference in the external benefits yielded by each (Bennett et al. 2011). If this is true, authorities should consider extending subsidy lengths for ecological forests or reducing those for economic forests. In other words, on granting farmers property rights that will permit them to respond to price signals, the government's role is to get those prices right (Bennett et al. 2011). Second, the difference in the net private benefits of planting economic and ecological forests is larger than the government expected. In this case, the government should lower the targeted share of ecological forest.

While permitting farmers to choose the types of tree increases the plants' survival rate, Bennett et al. (2011) found that when farmers chose which plots to retire, the survival rate is lower than when it is the village leader who chooses the plots. This result is perhaps surprising, as farmers with the right to select plots might have been expected to be more invested in tending them. Bennett et al. (2011) hypothesise that since the subsidies paid by the GfG were comparable to, if not larger than, the net yields from the retired plots, farmers may be willing to take risks, and convert more land than they can properly manage.

**Table 6.5** Variations in key variables with program implementation rights

| Plot-level comparisons | | Household has the right to choose what to plant? | | | Household has the right to choose the plot? | | |
|---|---|---|---|---|---|---|---|
| | | No | Yes | Ho: invariance with regard to autonomy (p-value) | No | Yes | Ho: invariance with regard to autonomy (p-value) |
| Tree/grass type | Economic forests | 58.8 % | 72.3 % | 0.02 | 62.8 % | 65.6 % | 0.18 |
| | Ecological forests | 32.5 % | 20.5 % | | 27.2 % | 29.9 % | |
| | Grass | 8.7 % | 7.2 % | | 10.1 % | 4.5 % | |
| Post-enrolment labor inputs on plot (labor days/ha) | 1st year | 155 | 273 | 0.01** | 192 | 209 | 0.73 |
| | 2nd year | 96 | 146 | 0.20 | 106 | 132 | 0.51 |
| | 3rd year | 65 | 145 | 0.08* | 107 | 81 | 0.57 |
| | 4th year | 65 | 101 | 0.15 | 96 | 68 | 0.26 |
| Post-enrolment cash inputs on plot (Yuan/ha) | 1st year | 256 | 544 | 0.04** | 282 | 512 | 0.10 |
| | 2nd year | 75 | 164 | 0.02** | 109 | 108 | 0.96 |
| | 3rd year | 98 | 139 | 0.32 | 124 | 100 | 0.56 |
| | 4th year | 154 | 155 | 0.99 | 103 | 207 | 0.21 |

Source: Bennett et al. 2011
Note: **$p<0.05$, *$p<0.1$

## Land and Tree Ownership

For both types of forests, land ownership is guaranteed during the contract period, and the planted trees or grasses also belong to the households that own the land. The turning point for collective forest management occurred when the government issued the "Decision on Some Issues Concerning Forest Protection and Forestry Development" in 1981. This decision included three major components (Démurger et al. 2009):

1. The stabilization of forest tenure through property certificates provided to owners,
2. The distribution of use rights to rural households on non-forested land (know as "family plots"),
3. The introduction of a forestry Contract Responsibility System that gave households land-use rights on collective forest lands (know as "responsibility lands") (Démurger et al. 2009).

At the household level, family plots and responsibility lands are the two main forms of forest tenure. Tenure is guaranteed for all land converted by the GfG, regardless of whether or not they are family plots. Since 1981 land can be inherited, and since 1998 farmers can also "transfer" (i.e. sell) their forestland to other farmers, through direct sale, auction or lease.[2] Only forestland can be transferred, not farmland or buildings. The name of the farmer who purchases the land is recorded in the local Forestry Office (for forestland), the Municipal Land and Resources Bureau (for wasteland) or any other department or office that used to own the land. The buyer is thereafter recognised as the new owner of the land.[3]

Since 1981 farmers also have greater control over land products, such as wood, that they can sell for a profit and retain the proceeds. Special regulations also exist regarding the trees planted through the GfG. Planted trees cannot be cut down during the period of compensation. When the cash and grain subsidies expire, those who converted their farmland to forests may, upon approval of the relevant departments, harvest the trees on their land, provided that such harvesting does not cause damage to the overall ecological system. However, household-level decision-making and management rights on trees are not fully guaranteed, because tree harvesting is still subject to the approval of local forestry bureaus. The Forest Law of 1984 also established a system of state-determined timber harvest quotas, which means that a household has to apply to the local government for a quota in order to cut trees on its land. The quota system is still in force today, and strongly reduces the degree of autonomy available to farmers regarding the sale of timber (Delang and Wang 2012). Thus, in practice, the government will continue monitoring and regulating tree felling.

Bennett et al. (2011) found that land rights do matter in the rates of survival of trees and grasses. Their findings were consistent with those of Grosjean and Kontoleon (2009), who found that greater land tenure security over enrolled land could increase labor inputs. In particular, using a Tobit model, Bennett et al. (2011) found that survival rates on private land were on average 23 % higher than on contract land (chengbao tian), which was auctioned off or allocated by village leaders for a fee. Similarly, their statistical analysis of 455 enrolled plots could not convincingly reject the null hypothesis that trees grown on private and responsibility land (i.e. collective forest land) have equal survival propensities (the null hypothesis carries a p-value of 0.058), with an estimated difference in survival rates of 19 %.[4]

---

[2] More specifically, people classified as "rural dwellers" in their household registration system (*hukou*) could sell it to other "rural dwellers".

[3] In some respects, China wants to continue considering itself a socialist country, and the private ownership of land is still considered anti-socialist. All land in China belongs to the government. There is no English word to reflect the kind of tenureship enjoyed by Chinese peasants, since people are, in fact, granted a range of rights that exceed the usual understanding of "tenureship". These rights have kept changing with the passage of time. For simplicity, in this book we will state that farmers "own" the land.

[4] Bennett et al. (2011) acknowledged that there are only six private plots in the sample, so that even though the large survival rate differences result in low p-values, larger datasets that may stratify on land rights and retirement status would be required before conclusions can be drawn on the effects of land rights. That the responsibility land variable is statistically insignificant could also be due to the noisy signal it provides regarding actual rights over a given plot.

Given that agricultural yields are substantially lower on contract and responsibility land, these results imply that retiring contract land (the omitted category) or responsibility land while also granting the household secure, long-term tenure on it, might do a great deal to boost survival rates without significant loss of agricultural production (Bennett et al. 2011). Bennett et al. (2011) acknowledged the possibility that contract and responsibility lands were lower in quality, in which case the regression results provided an upward-biased estimate of the impact on survival of granting private land rights. However, the authors contended that this omitted variable bias was unlikely to fully account for the different levels of output.

**Conclusions**

This chapter has reviewed issues related to the plants promoted by the GfG: economic trees, ecological trees, and grass, and discussed the survival rate of the trees. The survival rate of the trees is often well below the minimum stipulated for subsidy delivery, even though official data show a survival rate of between 90 and 100 %. However, one should recognise that even survival rates of between 60 and 70 % are relatively high. Most farmers prefer to plant economic trees, since they can obtain relatively high, annual benefit from the sale of non-timber products, such as fruits. If the trees stop producing fruits, the trees can be cut and the timber sold. Meanwhile, ecological trees can only generate limited incomes through the sale of wood from thinning, and the farmers have to wait many years before they can fell the trees and sell the timber. In addition, the sale of the timber is not guaranteed; farmers have to apply for a logging quota from the Forestry Bureau; they may have to wait several years before they are allowed to log their trees; or, they may receive a permit to cut only a small fraction of the total they applied for, which makes the logging uneconomical. In spite of this preference for economic trees, in most places the national standard of 80 % of the land being reforested with ecological trees is respected or exceeded.

# Chapter 7
# Household Selection, Participation and Attitudes

**Abstract** This chapter looks at household attitudes and engagement. In most places not all those who joined the program claim that they did so voluntarily, though many farmers were willing to convert their least productive land, especially when they had a surplus, and their remaining land was sufficient to grow enough food for subsistence. On the other hand, most researchers found that the Grain for Green is now a very popular program, since the funds are rather generous and the payments regular, and there is a visible improvement in the ecological conditions of the areas where it has been implemented.

**Keywords** Household participation • Household attitudes • Program targeting • Program impact • Voluntarism

## Introduction

The Grain for Green program directly engages millions of rural households as agents of project implementation, since the GfG is essentially a public payment scheme for environmental services. The number of participating households increased very rapidly once it was implemented nation-wide, from 3,577,296 in 2001 to 25,087,775 in 2005 (Ke 2007). By the early 2010s, Mao et al. (2013) states that a total of 30 million households and 120 million farmers were involved in the program. Thus, a large proportion of China's farmers benefited from the GfG, or live in villages in which someone participated in the GfG program. There is no doubt that the GfG has had a considerable impact on farming communities, even though it is clear that it is not the only project undertaken, and in many cases it interacts with other projects (Zhang et al. 2006). This chapter discusses farmers' attitudes about the GfG, the extent to which participation is voluntary or coerced, and the engagement of farmers in reaching project goals.



C.O. Delang, Z. Yuan, *China's Grain for Green Program*,
DOI 10.1007/978-3-319-11505-4_7

## Program Targeting and Impact

Wang and Maclaren (2011) examined the criteria of household selection into the GfG with a probabilistic model, and confirmed that households with lower average productivity were selected by the local governments. This means that they likely targeted the poorest households. On the other hand, Zhang and Liu (2006) directly analyzed the relationship between participation and poverty status to evaluate the impact of the GfG on poverty alleviation, and showed that the enrolment into the program was negatively related to the poverty rates, implying that poverty reduction was not a serious consideration when the farmers were selected.[1] These conflicting findings in different regions point to the difficulties in generalizing at the national level.

Uchida et al. (2007) argued that forestry officials designed the program to target sites that have a high potential to minimize the effects of off-site soil erosion, because the program criteria are not designed to target the poor directly. Nevertheless, poor households are the ones farming marginal plots in the upland regions of China's main watersheds, so the program is implicitly targeting these households. In Uchida's sample (2007),[2] participating and non-participating households share certain characteristics. For example, households in both groups were similar in terms of the number of children and adults as well as the age and level of education of the head of the household (Table 7.1). The two groups differ, however, in initial income level and asset holdings. Interestingly, before the program, participating households had 35 % more land than non-participants, but the land was 34 % further away from their houses and 32 % further away from sources of water (Uchida et al. 2007). This suggests that it was less productive, both in terms of person-day (since more time was required to go to the fields) and output per hectare (since in most cases water affects the productivity of crops). This resulted in participating households having 24 % lower incomes, living in houses that were 32 % cheaper,[3] and having consumer durables that were worth 39 % less (Table 7.1). It seems, therefore, that those who participated in the GfG had more land, but that it was less productive, which resulted in their greater poverty. With more land there was also greater opportunity to set aside a proportion of that land with the GfG.

To further investigate the factors determining participation in the program, Uchida et al. (2007) first estimate kernel densities to trace out the income distributions

[1] Zhang and Liu (2006) use data obtained from a unique panel survey conducted in 17 counties of North China by the MOF from 1998 to 2003. This survey was supplemented with village and county-level survey data. The 17 counties were randomly selected from 68 program-targeted counties in Hebei, Shanxi, and Inner Mongolia. Within the selected villages, a total of 188 households were sampled, with a total of 927 observations.

[2] Uchida et al. (2007) is based on a household survey carried out by the authors in 2003, and commissioned by China's MOF. The data set includes both participating and non-participating households. A total of 359 households in three provinces (Sichuan, Shaanxi, and Gansu), six counties, 18 townships and 36 villages were interviewed. Of these, 75 % of the households participated in the Grain for Green program (Uchida et al. 2007).

[3] Housing value is often recognized as a more accurate measure of wealth than income (since it is more easily observable and measured with less error) (Uchida et al. 2007).

**Table 7.1** Average household characteristics, income and asset holdings of participating and non-participating households in the GfG program

| | Participating households | Non-participating households | Difference in mean (t-test) |
|---|---|---|---|
| Number of households in sample | 253 | 86 | |
| Household size | 4.84 | 4.47 | 0.38* |
| Household head's age | 47.77 | 47.69 | −0.08 |
| Household head's educational attainment | 4.68 | 4.54 | 0.15 |
| Household landholding (mu) | 13.85 | 10.25 | 3.59** |
| Household landholding per capita (mu) | 3.05 | 2.54 | 0.52* |
| Weighted average of distance from each plot to house (m) | 1,029.47 | 769.92 | 268.56* |
| Weighted average of distance from each plot to water (m) | 1,068.59 | 810.99 | 257.59 |
| Weighted average of distance from each plot to road (m) | 852.33 | 814.17 | 38.16 |
| Income per capita (Yuan), 1999 | 1,404.41 | 1,850.41 | −446.00* |
| Value of house, 1999 | 13,659.45 | 20,066.54 | −6,407.10* |
| Fixed productive assets, 1999 | 842.80 | 948.47 | −105.67 |
| Consumer durables (Yuan), 1999 | 569.20 | 930.30 | −361.10* |
| Livestock inventories, 1999 | 384.58 | 414.39 | −29.80 |

Source: Uchida et al. (2007)
Notes
(1) *Significant at 0.1; **Significant at 0.05
(2) Housing value is based on self-reported values

of participating and non-participating households, using a fixed-effect logit model. The model includes county fixed effects to control for unobservable county characteristics. Uchida et al. (2007) used explanatory variables corresponding to three groups of factors that may have been used by local governments as criteria for deciding if villages and households should be included in the GfG program: environmental factors (slope, distance to waterway); wealth factors (income, land and asset holdings before the program); migration status before the program; and implementation costs (e.g. proximity to a public road), as well as other household characteristics. The authors used a random effect Tobit model to explain the total area of land in the program by each household. The covariates in the model are the same as those in the logit model except that the maximum slope of the household's plots is not included. This model is estimated with random village-level effects (Uchida et al. 2007).

The kernel density estimates of log of incomes per capita for participating and non-participating households illustrate that the income distributions of the two groups before the program (1999) were quite similar (Fig. 7.1). Although the means are statistically different at the 10 % level and the variance of participating households

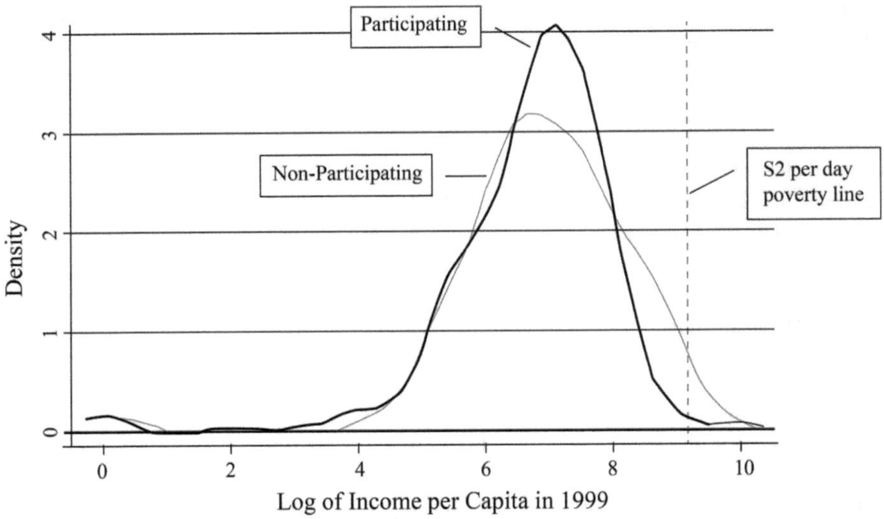


**Fig. 7.1** Kernel density of log of income per capita for participating versus non-participating Households in 1999 (Source: Uchida et al. 2007)

was somewhat smaller than for the non-participants, it is difficult to see striking differences (Uchida et al. 2007).

The results of the determinants of participation analysis also suggest that low income households were neither included nor excluded disproportionately from the GfG (Uchida et al. 2007). The estimated marginal effect of income level in 1999 was zero and insignificant (Table 7.2, column 1). Thus, holding all other factors constant, the poverty status of a household was not a determining factor in participation. The household's income level also did not affect how much cropland area was retired (column 2) (Uchida et al. 2007). Such a result would suggest that the program did not excessively target the poor. However, a more positive interpretation can be that the poor were not disproportionately excluded and that most people in the study areas were poor. Annual net income per capita in China in 1999 was Yuan 2,210 for rural households and Yuan 5,854 for urban households. As a comparison, this sample average was Yuan 1,518, which makes it well below the national average for rural households. Furthermore, a substantial number of relatively poor households were included (Uchida et al. 2007). We may assume that since the whole population in this study area was poor, the government did not further assess their level of poverty but treated them as a whole.

The results of the logit and tobit analyses of Uchida et al. (2007) illustrate other determinants of participation. For example, households with higher-sloped plots were more likely to participate (0.083), a finding that was expected since the steepness of a plot's slope was regarded as the principal criterion of site selection for the program (Uchida et al. 2007). The estimated coefficient on household head's age (0.008) suggests that households with an older household head was more likely to participate, a finding consistent with studies examining the determinants in the US CRP program

**Table 7.2** Fixed-effect logit and random-effect tobit regressions results explaining determinants of households' program participation and land size in the GfG

| | (1) Program participation (1 = participate) | (2) Area in GfG (mu) |
|---|---|---|
| Total agricultural revenue per capita in 1999 (Yuan) | –0.000 | –0.001 |
| Income per capita in 1999 (Yuan) | –0.000 | 0.000 |
| Land holding per capita (mu) | 0.010 | 2.150*** |
| Maximum slope among household's plots | 0.083*** | |
| Fixed asset value in 1999 (Yuan) | 0.000 | –0.000 |
| House value in 1999 (Yuan) | 0.000 | –0.000* |
| Livestock value in 1999 (Yuan) | 0.000 | –0.001 |
| Number of adult migrants in household | 0.099* | 0.370 |
| Distance from road to plots, weighted average (meter) | 0.000* | –0.001 |
| Distance from plots to house, weighted average, ln (meter) | 0.113** | 1.710** |
| Distance from plots to water, weighted average (meter) | 0.000 | 0.000 |
| Household size | 0.029* | 2.120*** |
| Household head's age | 0.008* | 0.018 |
| Household head's educational attainment | –0.004 | –0.210 |

Source: Uchida et al. (2007)
Notes
(1) Means of marginal effects are reported in column (1). In column (1) a constant is not estimated since county fixed effects are included
(2) *Significant at 10 %; **Significant at 5 %; ***Significant at 1 %

(e.g. Skaggs et al. 1994). Finally, the estimated coefficient of the variable measuring the number of adult migrants in the household (0.099) suggests that the probability of participating in the program was 10 % higher for a household with an additional adult migrant (or at least there was a positive correlation between migration and participation) (Uchida et al. 2007). For households that had already sent a household member into the migrant labor force before the program, the GfG program may have become an opportunity to take some cropland out of production, since many forestry activities result in labor savings compared to cultivation (Uchida et al. 2007).

Uchida et al. (2007) also tested whether or not the explanatory variables differed among the three provinces. In the logit model, they found that coefficients are statistically significantly different for two regressors: "distance to waterway" and "maximum slope". The marginal effect of "distance to waterway" is small but positive and statistically significant for Sichuan Province, while statistically insignificant for the other two provinces. Similarly, the marginal effect of maximum slope is positive and statistically significant for Sichuan Province but statistically insignificant for the other two. These findings suggest that participation decisions were systematically different for Sichuan Province (Uchida et al. 2007).

Liang et al. (2012) addressed similar issues, in focusing on the role of household composition, through a 2008 survey in Zhouzhi County (Shaanxi province). The county has a total area of 2,949 $km^2$, most of which is located in the Qinling Mountains, a natural boundary between northern and southern China. Zhouzhi County is one of the poorest counties in Shaanxi province, with Yuan 3,023 the average per capita income in 2005. Industry is forbidden and arable land is scarce in the Qinling Mountains (Liang et al. 2012). Cultivation of crops is common, although agricultural production is insufficient to satisfy needs. A number of environmental protection policies have been implemented in these areas, including the GfG in 2002, the NFPP, and the quota system for timber logging. These ecological and environmental protection policies are often compulsory and implemented by the central government from the top down (Liang et al. 2012). The GfG introduced economic trees, such as Cornus (a genus of fruit-bearing plants), walnuts, and peppers, which are sold to small retailers who come to the villages at specific times of the year. Walnuts and peppers are expected to generate income within 3 years, while Cornus generate income about 5 years after being planted (Liang et al. 2012).

Following previous studies, Liang et al. (2012) used the livelihood approach as an organizing framework to better understand the ways in which households adapt to policies (Fig. 7.2).[4] Figure 7.2 presents his analytical framework used to hypothesize about possible household responses to the program. As the basic unit of production and reproduction in most rural areas of the developing world, households pursue a livelihood strategy by allocating and organizing their resources in a variety of activities. The framework highlights the role of household composition and its direct influence on human capital and other household assets. Arrows indicate influencing factors.

Liang et al. (2012) compared five key assets of participants and non-participants: land, family size, education, social capital and access to credit (Fig. 7.3). It is difficult, however, to understand the impact of the GfG as the study was based on a survey carried out in 2008, several years after the program was implemented, and no attempt was made to understand the original conditions of the participant households. During these years, as Liang et al. (2012) partly acknowledge, the key assets that could change the least were family size (due to Chinese population policies) and education (measured as the family member with the most years of schooling). The other assets could change quite dramatically, in particular social capital, access to credit, and to a lesser extent land availability, since farmland and forestland can be bought and sold. However, Liang et al. (2012) hypothesize that "the differences

[4] Interviews were conducted in Zhouzhi county in April 2008. A total of 1,078 questionnaires were completed. In addition, 35 persons were involved in focus group and individual interviews. The survey covered both participating and non-participating households with a variety of detailed information on demographic characteristics, production and consumption activities, incomes and other livelihood outcomes, as well as some basic information on each family member. In particular, the questionnaire addressed households' assets, which did not change significantly, even after households participated in the program. Multiple level cluster sampling was adopted as the questionnaire survey method. At the household level, cluster sampling was used for the questionnaire survey in 20 villages from the four selected towns (Liang et al. 2012).

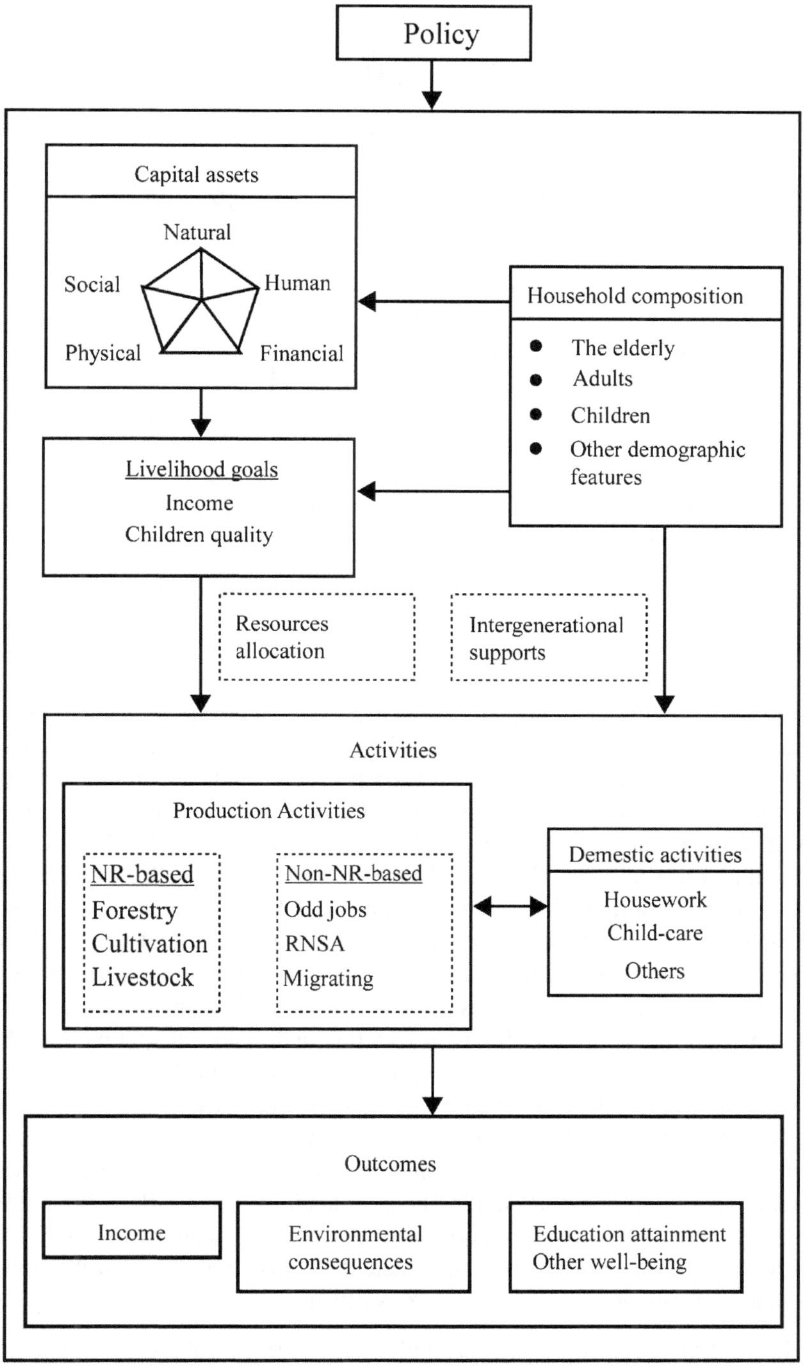


**Fig. 7.2** A livelihood framework with household composition (Source: Liang et al. 2012)

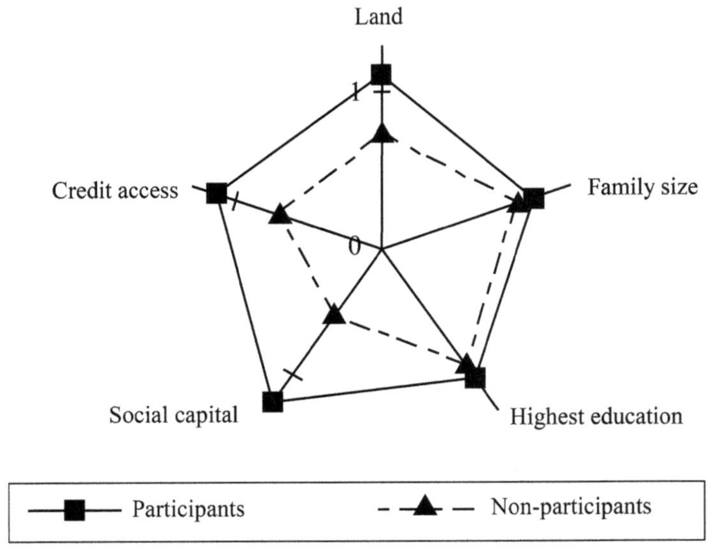


**Fig. 7.3** Differences in selected assets between participants and non-participants (Source: Liang et al. 2012)

between participants and non-participants may be explained as policy targeting and planning problems rather than policy effects on assets. Given the retired areas assigned by the government, households with more lands are more likely to be involved in the program" (p. 155). Liang et al. (2012) conclusions follows the findings of Uchida et al. (2007) discussed above and presented in Table 7.1.

Xu et al. (2006a), using the same dataset as Uchida et al. (2007), compared the incomes of both participating and non-participating households in Shaanxi, Gansu, and Sichuan, and found that participating households had slightly higher farming incomes than non-participating ones when they joined the program (in 1999), but much lower off-farm incomes, resulting in lower total incomes (Table 7.3). By 2002, the farming incomes of both participating and non-participating households had dropped compared to 1999. However, both animal husbandry and off-farm incomes increased; animal husbandry much more for participating than for non-participating farmers, and off-farm incomes by approximately the same percentage for the two groups: 30 % for participants, and 28 % for non-participants. Overall, the total incomes of non-participating households increased by 10 % from 1999 to 2002, while the total incomes of participating households increased by only 5 %. It is difficult to make sense of these aggregate data, since they may mask important differences among households. However, oddly, it seems that:

1. The program targeted households with higher agricultural incomes, which may seem at odds with its poverty-alleviation objectives.
2. The households targeted did have lower total incomes, but this was due to lower incomes from off-farm jobs.

**Table 7.3** A comparison of per-capita income changes induced by the GfG Program

| | | Non-participating households | | Participating households | |
|---|---|---|---|---|---|
| Income | | 1999 | 2002 | 1999 | 2002 |
| Farming | Pre-subsidy | 582.4 | 514.5 | 592.2 | 453.1 |
| | Post-subsidy | 582.4 | 514.5 | 592.2 | 525.3 |
| Animal husbandry | | 22.2 | 142.4 | 23.0 | 253.9 |
| Off-farm jobs | | 779.7 | 1,021.2 | 554.3 | 720.0 |
| Other | | 110.5 | 195.8 | 100.6 | 161.0 |
| Total | Pre-subsidy | 1,472.5 | 1,611.7 | 1,246.9 | 1,246.5 |
| | Post-subsidy | 1,472.5 | 1,611.7 | 1,246.9 | 1,313.2 |

Source: Xu et al. (2006a)
Notes
(1) The survey covered 84 non-participating households and 264 participating ones from six counties in Shaanxi, Gansu, and Sichuan
(2) The income figures represent real values after deflation with the 1999 consumer price index

3. Off-farm incomes of participant households were still lower than the off-farm incomes of non-participants, which is odd because the former are supposed to be "liberated" from (some) agricultural work, and able to engage in (more) off-farm work. The program might not have been very successful in shifting labor from on-farm to off-farm employment, which was one of the expected outcomes. Indeed, a survey in Wuqi County (Shaanxi Province) by Xue (2007) showed that only 30 % of the labor freed by the program migrated to off-farm work.
4. The differences in income between participating and non-participating households increased to the benefit of non-participating households, which would indicate that participation in the programs was not beneficial.

One must recognize that the survey was carried out in 2002, when the full impact of the program was likely not yet felt. Furthermore, as Xu et al. (2006a) write, without the GfG, the incomes of participants who were engaged in farming poorer quality, slope lands could have fallen even further behind those of non-participating households who were farming better land. Furthermore, as they caution, the GfG is not only (not even primarily) a poverty-alleviation program, and even if the GfG effect on income is small, its impact on environmental conditions could be significant.

## Is Participation Voluntary?

As discussed, the program has fairly efficiently targeted the areas and farmers it was meant to target. The next question is the extent to which farmers' participation was voluntary. According to GfG regulations, household participation should be voluntary (State Council of China 2002a). Scholars who have examined the issue, however, note that this is not always the case.

A number of scholars asked who decided which land to set aside. Uchida et al. (2007), Bennett (2008), Xu et al. (2006a), Xu et al. (2010) examined the question using a 2003 household and village-level survey of 360 households in 36 villages in the three provinces where the GfG was first implemented, located in the upper reaches of the Yellow River Basin and the Yangtze River Basin: Shaanxi, Gansu, and Sichuan.[5] Since these data are from the three western provinces where the GfG was first initiated on a pilot basis in 1999, they provide information about program implementation over a comparatively long period of time (4 years). Furthermore, the data include both participants and non-participants, thus allowing for examination of overall targeting of plots in the selected areas.

One discovery from the survey was the predominantly top-down approach toward implementation. However, the dataset reveals considerable variation among provinces in terms of the degree to which various levels of government were involved in the selection of areas to retire and of what to plant on retired land (Table 7.4). According to the dataset presented by Bennett (2008), in Shaanxi, there seems to have been more autonomy at the village and household levels, since to the question "Who decides how much area should be retired?", 50 % replied the "Households" and 16.7 % replied the "Village". On the other hand, in Sichuan 83.3 % replied the "Township" and only 8.3 % reply the "Household". Similarly, to the question "Who decides which plots to retire?", in Shaanxi 41.7 % said the "Household" and 33.3 % said the "Village" while in Sichuan 83.3 % named the "Township" and no one mentioned the "Household" or "Village". Finally, there was also more autonomy in Shaanxi regarding the vegetation to plant on the retired land, with 33.3 % of the interviewees stating that the "Households" decided, and 8.3 % stating that the "Village" decided, while in Sichuan 100 % of respondents stated that the "County" decided. Given these differences, it seems that there was considerable variation among provinces. In some areas the implementation of the GfG was a bottom-up endeavour, while in other areas it was top-down, with little (if any) autonomy for the households and villages. These results suggest that extrapolating from case studies like these is problematic and nation-wide generalizations make little sense.

Indeed, some researchers point out that most households join the program voluntarily, while other researchers note that voluntarism is mostly theoretical. Xu et al. (2010) used the same survey that Bennett (2008) had used, but present different information on various aspects of local implementation. Based on the Ministry of Forestry's plan, most if not all of the farmers in the sample should have been eligible to participate because, while emphasizing that highly sloped plots should have been targeted first, the plan also allowed some leeway in targeting lower-sloped marginal land that had an impact on the local watershed (Xu et al. 2010). The survey, however, as detailed in Table 7.5, revealed that only around 53 % of surveyed households felt

[5] The analysis uses a 2003 household and village-level survey conducted by the Chinese Academy of Sciences' Center for Chinese Agricultural Policy (CCAP). Two counties per province, three townships per county, two participating villages per township, and 10 households per village were randomly selected, for a total of 36 village surveys and 360 household surveys (Xu et al. 2010).

**Table 7.4** GfG implementation in 2003 survey village

| Question to village leader | Total (n=36) | Shaanxi (n=12) | Gansu (n=12) | Sichuan (n=12) |
|---|---|---|---|---|
| Who decides how much land area should be retired? (Percent of respondents) | | | | |
| Province | 0 | 0 | 0 | 0 |
| County | 25 | 25 | 25 | 25 |
| Township | 50 | 8.3 | 58.3 t | 83.3 |
| Village | 13.9 | 16.7 | 25 | 0 |
| Village small group | 0 | 0 | 0 | 0 t |
| Households | 22.2 | 50 | 8.3 | 8.3 |
| Other | 0 | 0 | 0 | 0 |
| Who decides which plots to retire? | | | | |
| Province | 0 | 0 | 0 | 0 |
| County | 16.7 | 16.7 | 0 | 33.3 |
| Township | 52.8 | 16.7 | 58.3 | 83.3 |
| Village | 36.1 | 33.3 | 58.3 | 16.7 |
| Village small group | 2.8 | 0 | 8.3 | 0 |
| Households | 16.7 | 41.7 | 8.3 | 0 |
| Other | 0 | 0 | 0 | 0 |
| Who decides what to plant on retired land? | | | | |
| Province | 0 | 0 | 0 | 0 |
| County | 66.7 | 41.7 | 58.3 | 100 |
| Township | 19.4 | 16.7 | 41.7 | 0 |
| Village | 5.6 | 8.3 | 8.3 | 0 |
| Village small group | 0 | 0 | 0 | 0 |
| Households | 11.1 | 33.3 | 0 | 0 |
| Other | 2.8 | 0 | 8.3 | 0 |
| Who conducts inspections and verifications? | | | | |
| County | 100 | 100 | 100 | 100 |
| Township | 100 | 100 | 100 | 100 |
| Village | 2.8 | 8.3 | 0 | 0 |
| Other | 97.2 | 91.7 | 100 | 100 |

Source: Bennett (2008)

that they could choose whether to participate (61.7 % of the participants and only 25.9 % of non-participants). The range was from 65.8 % of households in Shaanxi, to 45.5 % in Sichuan, and only 31 % in Gansu (Xu et al. 2010).

According to Tables 7.4 and 7.5, a majority of the surveyed farmers indicated that the government did not consult them prior to program implementation. Similarly, a majority of farmers affirmed that they did not have the right to choose which plots and how much of their cropland to retire, nor the right to select the tree/grass species to be planted (Bennett 2008). A case study of one program in Guizhou also notes that, in practice, rules dictated by government officials determined the species to be planted (Gong and Xu 2004). Similarly, a 2003 household and village-level survey

**Table 7.5** Farmer autonomy in GfG Program participation (N=345)

| Measure of autonomy | Percent that said "Yes" | | | |
|---|---|---|---|---|
| | Participants, all (n=264) | Shaanxi (n=103) | Gansu (n=85) | Sichuan (n=76) |
| Were the villagers asked their opinion about the project and how it could be best designed prior to the time that the project was implemented? | 42.8 | 41.7 | 41.2 | 46.1 |
| When your village began the GfG, did your household have autonomy to choose whether to participate? | 61.7 | 72.8 | 43.5 | 67.1 |
| Did you have autonomy in choosing the types of trees to plant? | 36.0 | 47.6 | 34.1 | 22.4 |
| Did you have autonomy in choosing which areas to retire? | 34.5 | 53.4 | 15.3 | 30.3 |
| Did you have autonomy in choosing which plots to retire? | 29.9 | 40.8 | 12.9 | 34.2 |
| | Non-participants, all (n=81) | Shaanxi (n=11) | Gansu (n=34) | Sichuan (n=36) |
| Could you participate in the GfG if you wanted to? | 25.9 | 45.5 | 29.4 | 16.7 |

Source: Xu et al. (2010)

conducted by the Chinese Academy of Sciences' Center for Chinese Agricultural Policy found that only 36 % of households (47.6 % in Shaanxi, 34.1 % in Gansu, and 22.4 % in Sichuan) stated that they had autonomy in choosing which type of tree to plant (Bennett 2008).[6] Similar conclusions were also reached by Yin et al. (2013). These findings regarding lack of household autonomy in participation choice runs counter to the program's explicitly stated principal of voluntarism.

Beyond this fundamental issue of choice, local people tend not to plant or maintain the trees and grass properly, if they are not consulted as to which land to set aside, and which plants to plant (Xu et al. 2006a). This may result in meagre survival and growth rates.

Uchida et al. (2005) also presented some evidence that a large portion of non-participating households wanted to participate but could not. Uchida et al. (2005) asked 87 non-participant farmers why they were not in the program. A majority (47 households) indicated that they did not participate because their sloped plots were not included in the program area. Only eight households indicated that they could have participated but did not want to (mostly because either the compensation level was not high enough or they did not believe that the compensation

[6] It is unclear whether this refers to the category ecological v. economic trees, or simply the species of trees.

would be delivered by the government). In sum, a majority of the households not enrolled in the program also had plots with steep slopes and appear to have wanted to participate if they had been given the opportunity in 2002 (Uchida et al. 2005). Xu and Cao (2002) went even further in concluding that enrollment in the GfG was completely involuntary.

The truthfulness of farmers' answers may be questioned. Wang and Maclaren (2011) suggested that the families who experienced a decrease in income were more likely to claim that the GfG was an forcibly imposed by the government.[7] On the other hand, as Bennett (2008) argues, even when participation is voluntary, the selection process is strongly influenced by China's rural government structure, since only households in participating villages are able to enter the program. Furthermore, as with many policies in rural China, the onus of actual implementation falls on village and township governments, which serve as the key mediators between the central government and rural households (Wang and Maclaren 2011). In other words, the selection of participants is based on whether land is within the program-targeted area. Thus, the GfG continues to adopt a top-down approach in choosing the targeted area and in determining participation quotas after the pilot phase (Bennett 2008).

## Households' Attitudes

In 2002 and 2003, soon after the implementation of GfG, Zhao et al. (2010) reviewed the cases of the Zhongyuan and Chashan small watersheds in Yunnan province. The sites are impoverished minority areas with a complex topography (high mountains, deep valleys, and great variation in elevation) and a distinct stereoscopic climate. Both watersheds experience an increasing impact of human activities on the land, including serious water erosion and soil loss, resulting in food shortage, and the reclamation and cultivation of the steeply sloping land. The Zhongyuan watershed is in the middle part of the Nujiang river valley with 617 households and a population of about 2,500. The Chashan watershed has a population of over 7,300, with 1,270 households. The GfG was introduced in the Zhongyuan watershed in 2001 and a year later in the Chashan watershed.[8]

[7] Wang and Maclaren (2011) found that the policy was welcomed by 60 % of the surveyed farmers, while the remaining farmers perceived government pressure to participate in the program. In contrast to farmers, county officials responded enthusiastically when the central government started the program, because of the fund counties would receive, and the top–down system of administration.

[8] Zhao et al. (2010) used field survey and semi-structured interviews to collect first and second-hand materials. More than 100 households were interviewed in 2002 and 2003: in the Zhongyuan watershed, Zhao et al. (2010) interviewed 48 randomly selected households, while in the Chashan watershed 57 randomly selected households were interviewed.

Zhao et al. (2010) found that 96 % of the surveyed farmers in Zhongyuan, and 80 % of those in Chashan, have a general understanding of the GfG and its relevant policies. Also, most farmers had a positive view of the program, because of the improved environmental conditions and the payments made. Some farmers, however, worried about the stability and continuity of the project and its policies. In particular, they feared that the government would stop paying compensation to the farmers, which would jeopardise its achievements. As a result, the farmers' actual participation varied significantly (Zhao et al. 2010):

1. Farmers with sufficient fertile lands in flat areas were more willing to participate in the project and to convert their steeply sloped lands into forestlands, especially when they had insufficient labor to farm all their land. Farmers who cultivated the marginal lands on the steep slopes were willing to participate in the program because the grain and cash subsidies provided by the government exceeded the yields from these lands.
2. Some farmers were willing to convert the dry lands at mid-high elevation but were reluctant to convert the paddy fields at low elevation, since cropping cultivation was their only source of food. If all the croplands were converted into forestlands but food and cash subsidies provided by the government were not sufficient, their livelihood could not be sustained. As far as the dry lands at the mid-high elevation were concerned, the grain output from these dry lands was generally below the GfG subsidy level, so the farmers were willing to convert that land into forestlands.
3. Farmers who engaged in animal husbandry were not willing to convert their croplands into forestlands, even marginal croplands. This was because they needed those lands to plant fodder for livestock. Furthermore, they found that the cash and grain subsidies from the government were much less than their income from corn and livestock (Zhao et al. 2010).

In Dunhua county of Jilin province Wang and Maclaren (2011) found that farmers thought family income was the most important concern, with water and soil erosion and forest protection coming second. In other words, most families wanted the government to implement measures to increase income and alleviate poverty. Once poverty had been reduced, the farmers were willingly to pay more attention to environmental problems. Global climate change was, in general, perceived by the farmers as an irrelevant matter in their lives and ranked last among their concerns.

Participation should have been based on a combination of self-selection by households and final selection by the local governments. This means that a compromise had to be found between the households, who wanted to select the land to set aside based on their expected returns, and the government, whose priority in selecting land was maximizing environmental benefits minus the opportunity costs of lost agricultural production (Uchida et al. 2005).

**Conclusions**

This chapter has reviewed the literature on the attitude of farmers toward the GfG. According to program regulations, participation in the GfG should have been voluntary, and no farmer should have been forced into it. However, the evidence shows that often farmers have not been able to choose whether to join the GfG or not. Also, it is not always the poorest farmers who joined the program, even though poverty reduction was one of the objectives of the GfG. On the other hand, one may argue that in the communities where the GfG was implement all farmers were relatively poor (by national standards), and therefore their individual poverty level might have been disregarded. All studies were based on surveys, and respondents' reliability is sometimes questionable; farmers who benefitted from the GfG may say that they volunteered, while farmers who did not benefit may say that they were forced to join.

# Chapter 8
# Planning and Implementation

**Abstract** Chapter 8 discusses the institutional context within which the Grain for Green was set up and operates, and the role of each level of government (national, provincial, prefectural, county, township) in its implementation. The implementation of the Grain for Green is complicated by the fact that China has traditionally had a very centralized political structure, with decisions being made in Beijing and little inputs from the regional and local governments. By contrast, the Grain for Green is a relatively decentralized program, with important decisions made at the grassroots level. The organization of such a large program, involving over 30 million households in 1,897 counties nationwide, is bound to face problems at the planning and implementation stages, and we review some of these problems in this chapter.

**Keywords** Bureaucratic reforms • Political reforms • Program planning • Five-year plans • Program implementation • Property rights

## Introduction

The role of the government is important for every reforestation program, specially when it is of the scale of the GfG. In China, the role of the government is compounded by the fact that China is a very centralized country, and the government exerts a great deal of control in most branches of the economy.

This chapter looks at different aspects related to the institutional organization of the forestry sector and the GfG. First, we discuss the bureaucratic reforms that took place within the central government's agencies responsible for forestry, relating it to the political and economic conditions of China after the liberation. Second, we discuss the Five-Year Plans, and in particular the increasing importance of forestry since the tenth Five-Year Plan of 2001–2005. Third, we discuss the administrative structure of the GfG, which is, in the Chinese context, a very decentralized program. Finally we discussed the implementation of the GfG, and some of the problems that arose.



C.O. Delang, Z. Yuan, *China's Grain for Green Program*,
DOI 10.1007/978-3-319-11505-4_8

## The Ministerial Reforms from 1949 to the 1990s

At the time of the liberation, the Chinese Communist Party (CCP) inherited a country which was politically fragmented, had weak political institutions at the national level, was very poor, and had very rudimentary infrastructure outside the main urban areas. Furthermore, and perhaps more importantly, it lacked qualified and experienced civil servants to manage and transform the country. While the Kuomingtang (KMT, in power from 1912 to 1949) had had more than 2 million civil servants, the CCP had at its disposal only some 720,000 qualified people to serve as civil servants in government administration (Macfarquhar and Fairbank 1978a: 74). In many government ministries, the number of civil servants was clearly insufficient. For example, in 1950, the Ministry of Forestry (MOF) had a staff of 27 people and two consultants from the Union of Soviet Socialist Republics (USSR), responsible for managing forest areas measuring some 100 million ha (the size of Saudi Arabia, Mexico or Indonesia) (Li 1988a; Wang and Delang 2011).

Although the new leadership had a decade of wartime administrative experience to draw upon, this experience was geared towards the administration of smaller regions, not the development of such a large country. At the same time, there was a need to reorganize the administration of the country to facilitate the reforms envisioned by the new leaders. To compensate for the lack of expertise and qualified personnel, China required foreign advice and technology. However, the West rejected the People's Republic of China (PRC),[1] recognizing instead the Kuomintang (KMT) government of Taiwan as the legitimate ruling government of China. Hence, China was forced to turn to the USSR for expertise, capital and technology (Wang and Delang 2011). However, the USSR was a very different country at the time of its Communist Revolution, so the experiences of the USSR were not always applicable to China. Because of the lack of valuable examples from other countries, China had to develop its own policies, and the institutions that would support such policies. The result was a trial and error approach, whereby China reformed its bureaucracy whenever a new format was deemed to be more effective. This section discusses the reforms that took place within the forestry sector.

During the period of national postwar economic reconstruction (1949–1958), the main goal of Chinese forestry was to provide timber for industrialization and large-scale construction. At the same time, forests were considered unused farmland, and logging was encouraged. To this end, the Ministry of Forestry and Land Reclamation was founded in 1949 (before 1949, the forests were managed by the Ministry of Agriculture) as an organization directly under the State Council.[2] It internally

[1] For example, the US reduced the export of timber to China from 399.78 million British pounds of timber in 1929 to 19 million pounds of timber in 1949 (People's Daily, 9 December 1949).

[2] In China, the State Council is in charge of the formulation of forestry policies and presents most forestry initiatives to the Standing Committee of the National People's Congress (NPC). The NPC is the highest body of state power and meets annually for about 2 weeks to review and approve major new policy directions, laws, the budget, and major personnel changes. The State Council, which is chaired by the Premier, is the highest administrative body of China and is responsible for carrying out the regulations and laws adopted by the NPC.

established the following branches: the Forestry Administration Department, the Afforestation Department, the Forestry Operation Department, the Forestry Utilization Department, and the General Office. Agriculture and forestry departments (*Ting* in Chinese) were established in each provincial government.

On 5 November 1951, only 2 years later, the Ministry of Forestry and Land Reclamation was split in two, creating the separate Ministry of Forestry while the land reclamation section was transferred to the Ministry of Agriculture. By 1953, the first year of the first Five-Year Plan in China, over 500 forestry-related agencies at the provincial or county levels had been set up nationwide. In 1954 a significant streamlining took place at different levels in the central and local governments. The State Council subsequently began to add agencies and offices, and by 1956 there were 81 units, the greatest number of government agencies since the establishment of the People's Republic of China (PRC) (Zhang 2008).

On 12 May 1956, the Standing Committee of the National People's Congress (NPC) decided to set up the Ministry of Forest Industry, which was responsible for administering nationwide afforestation, forest management and production of forest products through its ten departments and bureaus. It adopted the administrative system of the former USSR but with little effort to integrate with China's domestic situation. This resulted in an inevitable administrative dichotomy between forest harvesting and forest cultivation (Zhang 2008). Less than 2 years later, on 11 February 1958, the 1st National People's Congress decided to address that problem and merge the Ministry of Forest Industry and the Ministry of Forestry, and to set up a new Ministry of Forestry which had the responsibilities of the two former ministries (SFA 2000a).

During the Cultural Revolution (1966–1976), government departments experienced new, important reforms, initially with the aim of reducing the power of the central government. In 1970, the 79 agencies and offices that had been established were either dissolved or incorporated among 32 new offices, of which 13 were led by the military. The number of central governmental agencies dropped to its lowest since liberation (Zhang 2008). In June 1970, the Ministry of Forestry was disbanded and incorporated with five other organizations, including the Ministry of Agriculture and the Ministry of Agriculture Reclamation, into the Revolutionary Committee of Agriculture and Forestry. From 1971 these controls were relaxed, and in September 1971, the MOF held a new National Forestry Conference with the intention of addressing the problem of deforestation (Zhang 2008).

In early 1975, the MOF was renamed Ministry of Agriculture and Forestry. In May 1978, it was decided to make the Ministry of Agriculture and Forestry directly subordinate to the State Council, but administered by the Ministry of Agriculture on behalf of the State Council. Less than a year later, in February 1979, the Central Committee of the CCP and the State Council decided to revert to the previous conditions, and detached forestry functions from the Ministry of Agriculture and Forestry. The new Ministry of Forestry was again responsible for the management of the national forests and forest industries (Zhang 2008).

The reforms at the ministerial level during the three decades that followed liberation is indicative of the turbulent and chaotic political organization during these forming years, as the Chinese government was looking for a formula that worked in its socio-economic context. By the mid 1970s, the disruptions of the Cultural

Revolution and the Great Leap Forward ended, and the initial stages of the economic reforms were set in motion, including the reorganisation of government institutions. In December 1978, the 3rd Plenary Session of the 11th Central Committee of the Communist Party of China set in motion the "Reform and Opening Up" policy. From 1982, the State Council carried out political reforms at different levels, to harmonize the new economic realities with a new political organization. These reforms included the simplification of the administration, and the number of ministries, subordinate offices, and administrative bodies in the State Council was reduced from 100 to 61. However, the highly centralised administrative system supporting the planned economy remained intact and a full transformation of governmental functions was not realized (Zhang 2008).

The reforms of 1982 also affected the forestry sector. The Ministry of Forestry (MOF) was given more independence, as well as the responsibility to plan forest policies and the ability to better implement such policies with additional, better trained, staff and a greater budget. The MOF also undertook more research and development to increase the productivity of forests, or their economic benefits. Overall, however, the main objective of the forestry sector was still that of producing timber for the national economy, and the reforms did not stop the forest cover from shrinking. In particular, in the South, reforms in forest tenure resulted in the overharvesting of forestry resources in collectively owned forest areas. In 1984 the first *Forest Law* was promulgated (see Chap. 1) (Zhang 2008).

In 1997 and 1998 there were severe "natural" disasters – droughts and floods– on the Yellow and Yangzte rivers. These were blamed (partly) on the deforestation that had happened upstream. On 10 March 1998, the Ministry of Forestry was re-organised as the State Forestry Administration (SFA),[3] making it a body for specialized economic administration, and broadening the scope of its activities (Fig. 8.1). The work of the SFA involves the management, protection, and development of national natural resources (forests, wetlands, wild animals, and plants) and conservation activities (afforestation and reforestation programs, natural reserves, and forestry reforms) (Zhang 2008; SFA 2013a). Its primary responsibilities are related to managing the forests to recover their ecological and environmental functions, i.e. "the prevention of water and soil erosion, control of sandification through biological measures such as afforestation and grass planting, organizing and directing the management of forest land and tenure and analysis of forest land expropriation" (Zhang 2008).

Since 1998, forestry development has enjoyed its most productive years in history, with the initiation of six major forestry programs (Chap. 2), a considerable increase in investments by state and local governments, and a prevailing sense of optimism as to the future state of China's forests, specially considering the previous decades of degradation and deforestation (Chap. 1) (Zhang 2008).

[3]In our book, in line with other publications, we call it "Ministry of Forestry". But the official name is State Forestry Administration.

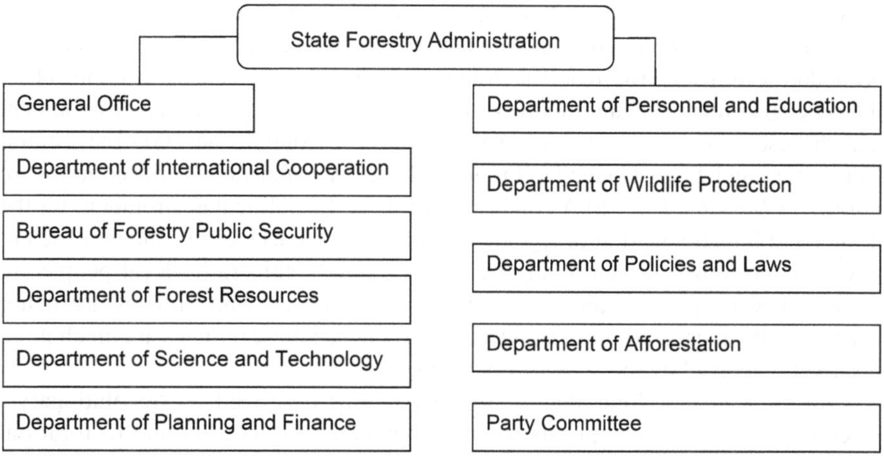


**Fig. 8.1** Forestry administrative structure in 1998 (Source: Zhang 2008)

## *Policy Framework: The Five-Year Plans*

Since 1953, the People's Republic of China has been implementing a series of Five-Year Plans that establish the blueprint and targets for national economic development. In a country where the state continues to exert tight control over much of the economy, the Five-Year Plans are key indicators of the directions and changes in development philosophy (Fan 2006).

The first Five-Year Plans completely ignored forestry, or simply encouraged the transformation of forestland into farmland, and the use of timber as a fuel or raw material for the manufacturing industry. As discussed in the previous section (see also Chap. 1), the first decades after the liberation were characterized by an unstable political and economic climate which prevented China from developing. However, the economic and political reforms from the late 1970s-early 1980s led to rapid economic growth, and by the late 1990s, two decades of economic reforms and rapid economic growth had changed China dramatically. It was no longer an undeveloped, collectivized, poverty-stricken country; it had become one of the world's most important economic centers. Unfortunately, the countryside was left behind during this period, and by the late 1990s, the central government acknowledged regional inequalities as a threat to social stability. Rural economic development and environmental protection became important parts of China's goal to build a harmonious society (Lai et al. 2013).

With the tenth Five-Year Plan (2001–2005) the government acknowledged the need for more active forest management and conservation. Indeed, in the tenth Five-Year Plan (2001–2005), the Chinese government included reforestation as one of the main tools for the nation's long-term development. On the environmental front, the plan set out to increase forest coverage to 18.2 %, and to take additional measures to protect and save natural resources (Pan 2006). In order to safeguard the environment from the rapid economic and social development that was expected to occur during the tenth Five-Year Plan (following China's accession to the WTO in 2001), the NPC also reviewed and modified the existing laws governing the environment and the use of resources (China Daily 2000). For example, in 2003 the Central Committee and State Council promulgated the *Decision on the Acceleration of Forestry's Development*[4] (SFA 2003d). Since then, the political environment for the design and implementation of legislation of forest policies has become more scientific (it is no longer driven purely by ideology) and democratic (local people are consulted when policies are designed and implemented, and policies are better adapted to the local conditions), and as a consequence these policies are much more positively received by the people concerned.

China's 11th Five-Year Plan (2006–2010), was described as "revolutionary", since it put particular emphasis to environmental conservation and restoration. At the same time, new policies were set up to tackle the problem of rural poverty, culminating in 2006 in a dramatic call by the national leadership to build a "new socialist country" and a "new socialist countryside", whereby the capitalist economic reforms that were introduced would be accompanied by policies to help maintain a more equal distribution of resources. China's 11th Five-Year Plan (2006–2010) included a call for increasing farmers' income, boosting agricultural productivity, and enhancing the natural environment (Zhang 2008). Almost all of the Plan's targets that were related to the natural environment focused on the conservation of resources. For example, the Plan required forest cover to be increased from 18.2 to 20.0 % (Fan 2006), a considerable progress from the tenth Five-Year Plan. Major ecological engineering projects were also pushed forward in the 11th Five-Year Plan period, to address soil erosion and other environmental problems (Jiang et al. 2013). That emphasis suggests that China's leaders were seriously reconsidering the country's development priorities. Indeed, in 2009, at the United Nations Climate Change Summit, President Hu Jintao promised the world that by 2020 China's forest area would increase by 40 million hectares, and that China will have 13 billion cubic meters of timber more than it did in 2005 (Xu 2011; Wan 2011: 113).

The emphasis towards a more sustainable economic development was further promoted in the 12th Five-Year Plan (2011–2015), in which GDP growth was no longer even officially considered a top priority, and the concept of ecological civilization was introduced. With this Plan, China's central government emphasised the need to change the mode of economic growth, and to develop a low-carbon economy, as two of its most important tasks. China's 12th Five-Year Plan clearly identifies the low forest cover as a constraint for the future development of the country, and the forestry sector was given a more prominent position (Wan 2011).

[4] The *Decision on the Acceleration of Forestry's Development* aims to optimize the structure of the forest industry, restore forestry resources, improve ecological restoration, harvest forest products in an efficient way, and increase farmers' income through a better management of the existing forests, reforestation, and the GfG program.

Some, such as Yeh (2009) question the genuineness of the desire of the central government to develop the west, arguing instead that concerns over the west were primarily due to the dependence of the more developed east on sound ecological conditions in the west, and perhaps the risk that ecological degradation in the west was jeopardising the nation's economic and political stability. Regardless of the motivations behind the government's emphasis to promote the reforestation of the country, and a "socialist countryside", it is clear that large amounts of money have been invested in the west, both for reforestation and economic development. Apart from the GfG, we also need to mention the *China Western Development* program,[5] planned from 1999 and in some ways closely related to the GfG, as the two programs complement each other. The *China Western Development program* was primarily a program meant to foster economic growth through the development of infrastructure (including transport, energy, and telecommunications). However, it complemented these purely economic goals with strong conservation and ecological restoration objectives, which included reforestation, and floods, droughts, and sandstorms containment projects (Lai 2002).

After this short introduction of the broad government policies and Five-Year Plans, in the following section we look more specifically at the GfG, and discuss the structure of forestry management. We then turn to the implementation of the GfG, and problems identified in the implementation of the program.

## Administrative Structure of Forestry Management

After the reform and opening up after 1978, China began to change from a planned economy to a market economy. With these reforms, the vertical administrative structure regulated by the central government gradually weakened, and local authorities have gradually had more independence (Lok 2009). However, that decentralization also brought confusion, and at times weakened the implementation of policies: delegating more power to local governments weakened the level of control of the central government and led to local authorities following the central government's instructions on the surface, without actually implementing them. To be successful, the GfG needed to achieve a balance. Here we look at how the GfG addressed that contradiction and managed to be relatively successful.

Due to China's large population and area, China's administrative divisions have been split at several levels since ancient times. Currently, there are five de facto levels of local government: the province, the prefecture, the county, the township, and the village. In 1954 when China's Constitution was written, the provinces became the primary administrative regions and were put under the direct control of the central government (Lok 2009). Below the province, there are the prefecture, the county, and the township. The power of the local government increases as the

[5]Also called "Go West", it mainly focuses on investment in infrastructure to help promote the development of, as well as the extraction of natural resources in, the West of China.

geographic area increases, and each level in the hierarchy is responsible for overseeing the work carried out by the level of government below it: the province oversees the prefecture, the prefecture oversees the county, and the county oversees the township. Because the village is at the grassroots level (usually a hundred or so households), it traditionally has no say in political decisions, including in some cases whether to join the GfG (The Central People's Government of the People's Republic of China 2013). In 2013, China had 34 provincial-level administrative units (23 provinces, five autonomous regions, four municipalities, and two special administrative regions), 333 prefecture-level administrative units (in 2004, 50 rural prefectures, 283 prefecture-level cities), 2,862 county-level administrative units (374 county-level cities, 852 county-level districts under the jurisdiction of nearby cities, and 1,636 counties), and 44,728 township-level regions (662 cities – including those incorporated into the four centrally controlled municipalities – 808 urban districts, and 43,258 towns) (The Central People's Government of the People's Republic of China 2013).

Chinese government policies are still very centralized, a legacy of the central planning approach of the past decades, and the same is true for most forest policies. In this sense, the GfG can be said to be a revolutionary program, as it involves various levels of government in its planning and implementation. Three levels of government are involved in planning the GfG: the central government, the provincial governments and the county-level governments:

1. At the central government level, the program is planned by the Ministry of Forestry (MOF) with the involvement of different departments, specially the Department of Western Region Development (WRD), and the Department of Development Planning (DP).
2. At the provincial and county levels, the Forestry Department and the Forestry Bureaus are able to comment and request particular changes, to adapt the program to the local conditions.
3. The county governments are responsible for coordinating its implementation. Normally, county forestry departments, in cooperation with township governments, conduct field surveys and report annual conversion plans to higher-levels, up to the MOF. The conversion plan includes the area to be converted by the GfG, the ratio of ecological and economic trees (no less than 80 % for ecological trees at the county-level), the selection of tree species and their distribution, the types of plantation, the supply of seedlings, and information about the management of plantations. After examination and approval by the State Council (in collaboration with the MOF, the WRD the DP), the plans are sent back to county-level governments for implementation.

China is politically very centralized, so the decentralization of power used to organize and run the GfG makes this a unique program from that point of view. The fact that provincial- and county-level authorities are able to adapt and adjust the program to the local needs and conditions, and that the participation of the farmers is theoretically voluntary, make the GfG less centralized than other Chinese government programs. However, it can still be said to be a top-down government program, since the State Council determines how much land is converted in each province, how much compensation the farmers receive, what kind of vegetation farmers can plant, etc.

## Implementation of the Grain for Green Program

The *Agricultural Technical Extension Law of 1993 of the People's Republic of China* and the *Circular on Stabilizing Forestry Extension System*, issued jointly by six central agencies, provide an enabling legal framework for forestry extension work. This is supported by financial inputs and capacity building by the government (Wu 2004). The task of forestry extension is carried out by extension bureaus of the forestry departments at different levels. At the national level, the Central Government has its Ministry of Forestry (MOF), which is responsible for laying down policies and guidelines for afforestation, forest management, forest industry, and other forestry development activities. The MOF has a specialized bureau responsible for extension work, and the Forest Departments at provincial, prefecture, and county levels have similar forestry extension divisions.

The forestry departments at the county level are the lowest-level forestry authorities. Extension personnel at the county level often serve the dual function of promoter (e.g. demonstrating planting techniques) and implementer (e.g. managing block plantations), as well as taking on the responsibility for forestry public education, publicity of the GfG, technical training (such as field visits), and individual outreach for new tree clones and new cultivation methods.

The main role of the township governments is to promote social and economic development (Wang 2013). Forestry stations at the township level are the grassroots units that organize and manage forestry production and protect forest resources. As such, the township government has forestry stations, normally with one to three extension technicians who are responsible not only for forestry extension but other administrative tasks. At present, there are 30,175 units at the township level with 151,101 employees, 87 % of whom are specifically trained to provide forestry extension services. Under the townships, there are village communities where farmer households are the basic production units. As a consequence, centralized management by the central government is integrated with decentralized management at the township and village levels (Wu 2004) (Fig. 8.2).

| Level of Government | Name of forestry organization | Body in charge of extension |
|---|---|---|
| Central Government | Ministry of Forestry * | Forestry Overall Station |
| Provincial Government | Forestry Department | Extension Division |
| Prefecture Government | Forestry Bureau | Extension Section |
| County Government | Forestry Bureau | Extension Section |
| Township Government | Forestry Station | |

**Fig. 8.2** Hierarchical structure of government and forestry organization and extension linkage (Note: *The official name of the Ministry of Forestry is State Forestry Administration (SFA), but in line with other literature on the subject, we retain the name Ministry of Forestry (MOF). Source: Based on Wu 2004)

Beyond government agencies, in recent years extension agencies have been set up within research organizations, such as universities, to extend their specific technologies. On the other hand, so far there are no NGOs or private extension agencies working among farmers. However, informally, stakeholders often undertake extension functions as well (Wu 2004).

The planning and implementation process for the GfG is complex because it targets millions of smallholders. It includes a broad set of government agencies and a great diversity of land-use types and technologies. The process of planning and implementation goes through the following steps (Li 2005).

Step One: The central government defines the overall scope, area and scale of the program. For example, how much money will be allocated in total for the GfG, which provinces may participate, what kind of land will be converted, which kind of vegetation can be planted, and how much money will be paid to farmers;

Step Two: The relevant provinces formulate provincial GfG plans (which include, for example, how much land the province may be able to convert, where such land is located, and how much land would be converted each year) and submit them to relevant central government bodies, including the MOF;

Step Three: The MOF examines and balances the plans of various provinces and then formulates the national GfG plan, which is then submitted to the State Council for approval;

Step Four: Once the national plan is ratified by the State Council, the MOF (jointly with other central agencies such as the State Development and Reform Commission) assigns tasks to the provinces and autonomous regions (for example how much land to convert each year in each province, with a given budget). The provinces or autonomous regions formulate a road map to carry out the various tasks;

Step Five: The provinces assign program tasks to lower-level government units, which in turn assign tasks to government units at even lower levels. For example, provincial authorities let prefectural authorities know how much land needs to be converted in each prefecture by which time. Prefectural authorities decide how much land each county needs to convert and inform counties accordingly;

Step Six: Local-level government units, normally county forestry departments in cooperation with township governments, conduct field surveys and delineate tasks to be implemented by farmer households, such as selecting the land to be reforested/afforested by each household;

Step Seven: Local level forest extension workers compile reports every year about how much land was set aside during that year, how much will be converted the following year, the level of subsidies to be distributed, which kinds of trees will be planted, etc. The higher-level authorities gather this information and integrate them into the following year's implementation plan. This is then reported up level-by-level to the MOF;

Step Eight: The MOF examines and approves the plans and sends them back, level-by-level, through the provincial, prefectural, county and township governments and forestry departments/bureaus;

Step Nine: Actual implementation takes place mainly at the local level. County-level governments send technical teams down to the townships; with the assistance of these teams, township governments organize farmers in the villages to implement the GfG;

Step Ten: The contract for the conversion of cropland is signed with the farmers' households, who are required to plant trees on their own retired croplands and the matching degraded lands (or barren lands).[6] To the extent that there is still quota for setting aside land through the GfG, more farmers can join;

Step Eleven: Inspections are conducted by authorities at various levels, including village, township, county, and provincial governments. The MOF also organizes random inspections;

Step Twelve: Farmers whose conversion work passes the inspections receive grain and cash compensation;

Step Thirteen: Government agencies at every level prepare work reports at the end of the year; at the same time, a work plan for the following year is prepared. The process is repeated yearly from Step Six.

The above steps can be summarized into three stages: planning and technical design, implementation, and inspection and evaluation. In a broad sense, the implementation process for the GfG is simply an extension program, through which the target population is provided with the necessary information, contacts, and services in order to understand and participate in the GfG. The extension has five main strategies:

1. Raise the public's awareness regarding the significance of the GfG;
2. Increase the farmers' understanding and willingness to participate;
3. Establish cooperation among government agencies at all levels, to facilitate the implementation of the GfG;
4. Include the farmers in the implementation process of the GfG;
5. Encourage technical transfer.

The following are some of the problems identified in the implementation of the program.

## *Funding Agency and Evaluating Agency*

The central government provides the greatest financial contribution to the GfG. Since the regional governments have few financial commitments, there is less motivation by the regional authorities to monitor whether the GfG is properly implemented by the farmers. For this reason, while the Forestry Bureau at the county-level is in charge of the evaluation of the program, provincial governments will conduct quality control, whose results may lead to rewards or punishments. Also, the MOF of the State Council will double check the quality control results, and conduct random inspections (Liao and Zhang 2008).

---

[6] If the farmers have the use right to barren land, they must plant trees on both the cropland and barren land.

## *Bias Towards Trees at the Expense of Grassland and Engineering*

A criticism of the MOF has been that it is biased toward planting trees, and has less emphasis on grassland and engineering improvements. Indeed, the GfG does not give sufficient consideration to the ecological and economic functions of grasslands in semi-arid areas and the need to restore these ecosystems. Partly because of this, the central government decided to create a separate program of Yuan 20 billion, under the coordination of the Ministry of Agriculture, to recover and revegetate grasslands in the west during 2003–2007. Additionally, while the MOF has been charged with administering the GfG, other government agencies responsible for agricultural and livestock production, water and soil conservation, poverty alleviation, and environmental protection are not formally involved. Therefore, inter-agency cooperation and coordinated implementation is very weak (Xu et al. 2006a).

## *Too Rapid Implementation*

Uchida et al. (2007) lamented local governments' rapid decision to implement the GfG, and their lack of transparency in the details of implementation, as officers avoided conducting interviews with potential participants at the onset of the program. The same problem was reported by Hori and Kojima (2008) in Y village, Mizhi County (Yulin City, Shaanxi Province). Using interviews with an official of the Mizhi Forestry Bureau, Hori and Kojima (2008) discovered that only 2 months passed between the time the County Forestry Bureau decided how much land each Township had to set aside, and the time farmers were instructed to plant trees and grasses on their land. After farmers were instructed to plant trees and grasses on their land, only a further 4–7 months passed until County Forestry Officers surveyed the fields to assess whether the farmers had complied (Table 8.1) (Hori and Kojima

**Table 8.1** Calendar of GfG in the Mizhi County (Yulin district, Shaanxi Province)

| Month | Progress |
|---|---|
| February | The Mizhi County Forest Bureau decides how much area is to be converted annually in each township |
| Early March | The township government decides how much area is to be converted in each village |
| Middle March | The leader of the Villagers' Committee allocates the area to be converted to farmers |
| End of March–April | Farmers plant trees and grasses |
| August–October | The Mizhi County Forest Bureau investigates the planted trees and grasses |

Source: Hori and Kojima (2008)

2008). The rapidity in selecting the land to be set aside and the farmers to be involved, raises the question of whether the most appropriate fields (the ones with the lowest productivity) and suitable farmers (the poorest) were selected carefully, or whether selection was done by broad strokes on a map. Indeed, the rapidity of selection might help explaining why not always the poorest farmers and the worst land was set aside (see Chaps. 5 and 6).

## *Lack of Flexibility*

While the direct engagement of households as core agents of program implementation sets a new direction in managing China's strained natural resources, several features of the program hark back to policies and mindsets of decades past (Yin et al. 2005; Bennett 2008). They include the top-down, simplistic contractual structure, the lack of sufficient consultation with local communities and rural households to identify their needs and constraints, and the campaign-style mass mobilization aimed at reversing, in one decisive thrust, a variety of adverse environmental outcomes (Yin et al. 2013). Instead of taking the traditional top-down approach, Liu et al. (2008) proposed that more input and feedback from local people affected by the policies should be actively sought and incorporated into the decision-making process. Moreover, according to Xu et al. (2006a) the central authorities have failed to realize the importance of the incentive structure, placing too much reliance on administrative campaigns, and not enough on contracts, open bidding, and other market-based mechanisms to carry out specific activities.

## *Fiscal Burdens to Local Governments*

Since the 1980s, decades before the implementation of the GfG, China experienced unprecedented economic reforms, also called "socialism with Chinese characteristics". A notable development was the decentralization of state control, leaving local provincial leaders to experiment with ways to increase economic growth and privatize the state sector (Brandt et al. 2008). Township and village enterprises, firms nominally owned by local governments but effectively private, began to gain market shares at the expense of the state sector. Local government revenue significantly increased. However, to some extent, China achieved high economic growth at the cost of increasing income inequality. According to the World Bank (2014), the Gini coefficient of China was 0.426 in 2002 (up from 0.371 in 1997), which is slightly above the internationally-recognized danger level.

Partly because of the increasing inequality, in 2000 the government started a financial reform for agricultural products (this reform was implemented gradually, and covered the whole national territory only in 2006). Through this reform, taxes on agricultural products, including trees and timber, were abolished. Since 2007, farmers in China no longer have to pay taxes or fees for agricultural or non-timber forest products, though they still pay limited fees for timber (Tao and Qin 2007). Abolishing agricultural taxes helped improve farmers' income and narrow the gap between the people living in the rich and the poor provinces. However, it also resulted in increasing the financial burden of many local governments. Prior to 2000, much of the income of local governments came from taxing farmers' agricultural output. Not surprisingly, setting aside agricultural land through the GfG, and engaging in the agricultural tax reform, led to a reduction in tax revenue to the local governments. In Kangding County of Sichuan Province, for example, local government income decreased by 28 % to Yuan 15 million during 1999–2001 (Dong 2003). At the same time, the GfG was organized in such a way that the expenses necessary for its implementation, such as the costs of extension work and of delivering seedlings and rice, had to be covered by the local governments. These changes resulted in the GfG constituting a great burden to the local governments, specially in the poorest areas (Liu and Zhou 2005). Only after mounting complaints and delayed execution and inspections, did the central authorities make adjustments by assuming the bulk of the administrative expenses (Bennett 2008).

### *Problems in Payment Delivery*

In some cases, the fiscal burden to the local government also resulted in delays of payments to farmers. Xu et al. (2004) and Bennett (2008) reported that in many regions compensation payments have not been (either completely or partially) delivered to their rightful recipients, due to delays and shortfalls in the payment of compensation (see Chap. 4). The early shortfalls in delivered subsidies are in part symptomatic of a key design flaw: poor administrative budgeting (Yin et al. 2013). Program coordination, inspection, and subsidy delivery for millions of plots is burdensome and costly, and yet the initial plan mandated that local governments bear these costs, while they were facing reduced revenue from taxation. Many counties in western China, which were designated as priority sites for cropland retirement, faced severe budget constraints, which impeded them to complete the payments to farmers (Bennett 2008).

### *Cost of the Program*

The GfG also created a potential fiscal burden for the national government, since it required a tremendous amount of funding. The largest share of the budgetary outlay is used to compensate farmers for setting aside their cropland and planting seedlings

**Table 8.2** Actual compensation for total area under grain for green program in Ningxia and Guizhou, 2000 (Yuan)

| | Ningxia | Guizhou |
|---|---|---|
| Actual compensation for program plot[a] | 137,942 | 21,364 |
| Amount of over-compensation | 75,557 | 1,994 |
| Amount of under-compensation | 24,063 | 6,603 |

Source: Uchida et al. (2005)

[a]It is assumed that the farm households in the survey were fully compensated for their program plots

on their land (Uchida et al. 2005). In some areas, the payments farmers receive in compensation are higher than the value of the crop they would be able to produce on the land, so there are clear opportunities for lowering the level of payments. Indeed, Uchida et al. (2005) identified the problem of poor targeting in terms of over-compensation, and large variance in compensation. Over-compensation, occurs when actual payment through the GfG are greater than the potential net revenue from growing crops on that land, while under-compensation occurs when actual payment are smaller than the opportunity cost of the land. Uchida et al. (2005), using a sample of households in Ningxia and Guizhou (see Chap. 4), found that if officials had compensated farmers at levels equaling the plots pre-program net revenue, they could have reduced expenditures by 60 % in Ningxia, while they would have had to increase expenditures by 18 % in Guizhou, to eliminate under-compensation (Table 8.2). Uchida et al. (2005) concluded that better targeting could have reduced the cost to the government, as well as to the farmers, by including non-program plots that had lower net revenues instead of the relatively more profitable program plots.

On the other hand, by paying farmers more than the opportunity cost of the land, and providing a direct transfer of capital from the central government to the farmers, the program has been very successful in raising the incomes of farmers. Since poverty alleviation is one of its objectives, the excessive payment of the program in some areas is not necessarily a problem, as it helps reduce inequality among regions. The direct payment to farmers is another characteristic that makes the program unique in China. Most other programs undertaken in rural areas are supposed to lead to the better standards of living, through such things as the development of infrastructure, or the improvement of irrigation or soil conditions, but do not include direct transfers of funds to farmers, so their impact (specially among the poorer farmers) may be ambiguous.

Beyond considerations of the opportunity cost of the land, some question whether such generous payments are necessary. Yin et al. (2013) for example, argue that instead of moving land management from one extreme (extensive cultivation) to another (extensive forest), alternative, more practical, scenarios can be construed by merging ecosystem services with economic/livelihood services. Evidence indicates that many communities, against the directive of the authorities, have adopted agroforestry regimes by continuing to grow annual crops and planting more commercially valuable trees on retired farmland (Yin et al. 2005).

## *Unfairness in Compensation*

From 2000 to 2003, the farmers were compensated in grain for the land they had set aside. From 2004, they were compensated in cash, for an amount that in 2004 corresponded to the average value of the grain that could be grown on their land (Yuan 3,150 per ha in the Yangtze River watershed and Yuan 2,100 per ha in the Yellow River watershed). Since 2008, this compensation has been halved. However, since 2004 the price of grain has increased, which has resulted in a considerable drop in the amount of grain they are able to buy with the cash they receive. For example, Ma and Fan (2005) estimated that in Xiqu Township, Minqin County (Gansu Province), farmers lost Yuan 3,852–4,000 per ha, partially because of increased prices for agricultural products since 2004. A further problem is that current payments for the two programs are relatively uniform across space, although there are large regional variations in the costs of program implementation. Xu et al. (2006a) argued that it is unclear why the standards, durations, and specific restoration measures could not be localized and made more flexible, and why communities' interests and stakes were not better incorporated into the program design and implementation.

## *Rights to Land and Land Products*

The success of the GfG does not only depend on how well the program is organized. Other relevant policies in agriculture, land rights, investment in infrastructure, and the broader economic development of the country are also essential to its success. In particular, the rights associated with land and the products of land have important implications for investment in agriculture or forestry (Brandt et al. 2002).

Restrictions on how the land can be used, for example rules limiting farmer's crop choice or the ability to convert land to alternative agricultural uses, may affect land productivity and income earned from farming. For example, restrictions on property rights, and in particular tenureship insecurity, may discourage investment that increases the long-term productivity of the land, thereby negatively affecting output. In the case of the GfG, insecure property rights may discourage care in the trees, which may affect their survival rate. For example Bennett et al. (2011) found that there is a close relationship between the property rights households hold over retired plots (in terms of being able to chose whether to retire the plot and what kind of plants to plant) and the amount of labor-time they were willing to expend, the amount of cash they were willing to invest, and the survival rate of trees.

China's rural economic reform also radically altered land tenure. With the introduction of the Household Contract Responsibility System (HRS, adopted nationwide in 1982) and the extension of land-use rights to households, agriculture shifted from a collective-based to a family-based economic enterprise. However, land was not privatized. Land ownership remains “collective”, with local governing bodies and officials at the village level exercising a major influence over household land allocation and land use (Dong 1996). Indeed, the formal ownership of farmlands lies with the local government, although farmers have use rights for at least 30 years (Wu 2004). The lack of formal ownership of the land is likely to remain unchanged for some time, because as Rozelle et al. (2001) argue, poorly developed credit markets, the lack of a land registration system, and an incomplete legal system make privatization of land at the current time inefficient, if not socially dangerous. However, the GfG has led to some reforms in the land tenureship system. Since trees need some time to grow and provide the best economic benefits, and then provide these benefits for many years, it was felt that 30 years was not sufficient, so the farmers were given the right to manage the land and keep the benefits from the sale of the timber or non-timber products for 70 years (State Council of China 2002a).

A survey of 274 villages by the State Statistical Bureau in 1992 provides an estimate of the percentage of land of the three most common tenure types. In most villages, land rights can be divided into private land[7] (*ziliu di*, around 6 % of the land in the 1992 survey) and collectively controlled land (*jiti di*, more than 90 %) (Brandt et al. 2002). Generally, leaders do not intervene into decisions on private plots and farmers enjoy a high degree of tenure security (Bennett et al. 2011). On the other hand, collectively controlled land involves various types of restrictions and obligations. Collective land includes three different tenure forms:

- ration land (口粮田, *kouliang tian*), which goes to farmers mainly to meet household subsistence requirements with no tax obligations;
- responsibility land (责任田, *zeren tian*), which is distributed to farmers under the condition that they deliver a low-priced grain or cotton quota to the state;
- contract land (承包田, *chengbao tian*), which is auctioned off or allocated by village leaders for a fee (Bennett et al. 2011).

Officially, ration and responsibility land have the most secure tenure among these collective land rights categories. However, significant variations exist in the de facto rights at the village level (for example, regarding tenure length, use rights and transfer rights), to the effect that the distinction between them is often somewhat blurred (Brandt et al. 2002).

[7] All the land in China belongs to the state, so ‘private’ land is not to be taken literally. There is no English word to represent everything that the farmer is allowed to do with this land. In general, households have almost complete control over the short-term and long-term management of the private plots, which at present includes usufruct, renting, selling and passing on ‘ownership’ to the children.

**Conclusions**

This chapter has reviewed the changing institutional arrangements that have governed the forestry sector, and in particular the organization of the GfG program. The GfG program, by its sheer size and number of people affected, is very difficult to organize, and will be prone to different kinds of problems at various levels. What complicates the execution of the GfG is that many of the areas in which it is implemented have a relatively weak local government, rather rudimentary infrastructure, little financial revenue, a very poor road network, and in general few options open to the farmers. Considering these constraints, and given the economic, social, environmental and political constraints that exist in rural China, the GfG can be said to have been a successful program. The GfG also stands out as a relatively decentralized program in the Chinese context, and as such is somewhat revolutionary. Having said this, it is still planned by the central government, which makes the broad decisions, including for example determining the ratio between ecological and economic forests, which trees are categorized ecological and economic, how much land to convert in each province, and how much to pay and for how long. It can also be criticized for not adapting sufficiently to local conditions, in terms of the choice of vegetation it promotes (for example by promoting trees in arid areas where rainfall is insufficient), or by having payments that are too uniform across the watersheds of the Yangtze and Yellow Rivers. Nevertheless, by engaging the political authorities at lower levels – from the province, prefecture, county and township – in the planning process, and adapting the policies to the feedbacks received, it can be said that an effort has been made to involve the local authorities in the planning and implementation of the program.

# Part III
# The Impact of the Grain for Green Program

As a reforestation program destined mainly for rural villages, the GfG has had both ecological and socioeconomic impacts. In the following chapters we review these impacts of the GfG in some detail. In Chap. 9 we review the ecological impacts, focusing on soil resources, water balance and carbon sequestration. Overall, the ecological impacts of the GfG are deemed to be beneficial, even though the nationwide impact of some factors, such as soil and water conservation, is difficult to ascertain because of lack of countrywide studies. Although the GfG is primarily an ecological restoration program, more research has been done on its socio-economic impact than on its ecological consequences. In part, this might reflect concerns about the program's impact on the affected population. Other factors, however, may be at work, including the interest of researchers and the sensitivity of ecological data to the Chinese government, which result in greater difficulty in finding information and obtaining permits to do fieldwork.

Chapter 10 looks at the impact of the GfG on the grain output and price. Between June 2002 and June 2004, grain prices increased, which was blamed on the reduction of farmland set aside by the GfG. As a consequence, the GfG was scaled down after 2004. However, most researchers agree that the GfG was not responsible for the increasing price of grain, because most of the land converted by the GfG was not very fertile. Furthermore, farmers could concentrate their efforts on their more fertile parcels, with the result that grain production overall fell only very marginally in GfG-converted areas. Instead, researchers argued that the drop in grain output, and subsequent increase in grain prices, was more likely due to the loss of farmland caused by urbanization in the eastern provinces (where the GfG was not implemented), as well as the shift from grain production to the production of other crops.

Chapter 11 discusses the impact of the GfG on the sources and level of income of farmers. While in many places agricultural incomes tended to dominate before the GfG was introduced, by relieving farmers from agricultural work, the GfG has had a considerable impacts on the economic structure and potential sources of income. With the GfG the income structure diversified, to include agriculture, GfG subsidies, the sale of GfG-sponsored forest products, off-farm work in the villages of residence, and migration. In terms of the incomes from GfG-induced land use changes, a distinction has to be made between economic trees, ecological trees, and grassland. Researchers agree that economic trees bring higher profits to the farmers, but even among economic trees, not all trees bring profits comparable to crops, once the subsidies are excluded from the calculation. Most researchers have looked at the benefits per hectare. However, equally important are the benefits per person-day. When these benefits are considered, economic trees fare much better, because they require little manpower compared to crops. Whether benefits per hectare or per person-day are more important is open to debate. However, considering that in many rural communities most people of productive age have migrated to cities, either in their province of origin or in the eastern provinces, and there is insufficient labor available to farm crops on the set-aside land, the importance of benefits per person-day as a unit of analysis should not be underestimated. When considering benefits from labor input, many economic trees generate higher incomes than crops, and may be of great help to those who remain behind.

Chapter 12 reviews these issues in more detail, by looking at labor force redistribution. The GfG was expected to increase migration and off-farm employment. However, researchers found that in some areas farmers who joined the GfG program migrated less. These farmers were able to improve the productivity of their remaining fields (as well as farm fields of those who had left), which resulted in higher incomes and removed the need for migration. Successful migration also depends on qualifications and social networks, and it is likely that those who joined the GfG, often the poorest members in the villages, had fewer opportunities to migrate that their wealthier neighbors. In any case, the evidence shows that there is considerable variation among communities in terms of the impact of the GfG on income levels, and few generalizations can be made.

In Chapter 13 we examine the sustainability of the GfG. The GfG was originally scheduled to end in 2007, but was extended, and the subsidies are now set to end beginning in 2015 for the land that was first set aside, and later for other land. The question is whether the farmers will continue with the GfG-induced land use changes, or will revert back to the pre-GfG land uses once the subsidies end. There are constraints on cutting the trees, in particular a quota system, whereby the farmers need to obtain permission from the Forestry Bureau to fell their trees. Nevertheless, if the income from tree products do not compare favorably to those from cash crops, when the subsidies end there will be considerable pressures on the forest. The hope is that the rural economies have sufficiently transformed (through the GfG and other programs), and that off-farm opportunities abound, so that farmers no longer need to revert their least productive land to pre-GfG land uses. One issue that complicates assessments of sustainability is the fact that most studies were done during the first years after the program was implemented, when the monetary benefits from the economic trees could not yet be fully ascertained. However, some studies did try to estimate future changes in prices, and predict farmers' adaptation to such changes.

# Chapter 9
# Ecological and Environmental Impact

**Abstract** In this chapter we review the ecological impact of the Grain for Green. Because the Grain for Green is primarily a reforestation and ecological restoration program, the success or failure of the Grain for Green depends in large part on its ecological impact. Its ecological impact can be assessed using such indicators as the amount of land converted and afforested, changes in vegetative cover, water surface runoff and, very importantly, soil characteristics. Unfortunately there is no nation-wide assessment of the ecological impact of the Grain for Green, so it can only be gauged from case studies in selected regions. Most studies concur that the physical properties of the soil, including soil fertility, porosity, and nutrients, have improved, and soil erosion and river sedimentation have slowed down. However, most researchers agree that the impact of the Grain for Green in arid areas has not always been positive, given its emphasis on trees, rather than shrub or grass. Another important environmental impact of the Grain for Green is that of the amount of carbon sequestered by soil and trees. Unlike for the ecological impact on vegetation, water and soil, there are province- and nation-wide studies done on the impact of the Grain for Green on carbon sequestration.

**Keywords** Ecological impact • Soil fertility • Soil porosity • Soil nutrients • Soil erosion • Sedimentation • Water balance • Water runoff • Carbon sequestration

## Introduction

Because the Grain for Green is primarily a reforestation and ecological restoration program, the success or failure of the GfG depends in part on its ecological impact. Its ecological impact can be assessed using indicators that are immediately observable, such as the amount of land converted and afforested, changes in vegetative cover, water surface runoff and, very importantly, soil erosion (Table 9.1). Based on these indicators, most researchers agree that the program has been successful (except, according to many researchers, in the arid northern region), even though unfortunately there is no nation-wide assessment of the ecological impact of the GfG, so it can only be gauged from case studies in selected regions. In the following pages we review case studies carried out in Guizhou, Hubei, Shaanxi, Gansu and Hunan Provinces. These studies concur that the physical properties of the soil,



C.O. Delang, Z. Yuan, *China's Grain for Green Program*,
DOI 10.1007/978-3-319-11505-4_9

**Table 9.1** Indicators of ecological changes through GfG-led vegetation restoration

| | |
|---|---|
| Conservation of water resources | Vegetation coverage |
| | Annual runoff coefficient |
| | Land water storage |
| | Area ratio of the soil erosion |
| Conservation of soil and water | Soil erosion modulus |
| Soil amelioration | Soil bulk density |
| | Soil porosity |
| | Amount of organic matters |

Source: Yang (2005)

including soil fertility, porosity, and nutrients, have improved, and soil erosion and river sedimentation have slowed down. However, most researchers agree that the impact of the GfG in arid areas has not always been positive, given its emphasis on trees, rather than shrub or grass. Another important environmental impact of the GfG is that of the amount of carbon sequestered by soil and trees. Unlike the ecological impact on vegetation, water and soil, there are province- and nation-wide studies done on the impact of the GfG on carbon sequestration. The first part of this chapter looks at soil fertility and the conservation of water resources, as well as the impact of the GfG on soil erosion in arid areas. The second part of the chapter looks at the changes in carbon sequestration brought about by the GfG.

## Conservation of Soil and Water Resources

### *Soil Characteristics, Soil Erosion and Water Runoff*

Luo et al. (2003) examined changes in Boluo Village of Qingzhen Town (Guizhou Province) over 3 years. The study area is influenced by a humid subtropical monsoon climate and has an annual temperature of between 14 °C and 16 °C. The annual precipitation is 1,100–1,200 mm. In total 50 soil samples were collected from land with 15°, 25° and 35° slope, with and without GfG reforestation. Not surprisingly, the researchers observed that runoff became more severe when the degree of land slope increased. Reforestation reduced the loss of soil nutrients due to runoff, and let to a recovery of soil fertility: organic matter, nitrogen (N), phosphorus (P) and potassium (K) all increased after the GfG was implemented, although there was no significant change in the pH value (Table 9.2). Also, the permeability coefficient of saturated soils increased soil hold capacity by 1.66 times after reforestation, compared to the soil with no reforestation (Table 9.3).

Although satisfactory effects were observed and tested in soil conservation, Mei and Xiong (2003) questioned the performance of the GfG in water conservation. Luo et al. (2006) studied the physical properties of the soil on which four different species were planted through the GfG: *Triploid Chinese white poplar*, *Alnus cremastogyne*, *Cunninghamia lanceolata*, and *Hybrid Bamboo*. The ability to conserve

**Table 9.2** Amount of Yellow Soil Nutrients with and without the GfG

| Land use | | pH | Organic Matter (k/kg) | N (k/kg) | P (mg/kg) | K (mg/kg) |
|---|---|---|---|---|---|---|
| Without GfG | Average | 5.46 | 24.76 | 1.42 | 618.5 | 105 |
| (n=10) | Std | 0.87 | 13.83 | 0.73 | 165.8 | 41.2 |
| With GfG | Average | 5.38 | 26.46 | 1.62 | 725.8 | 110 |
| (n=13) | Std | 0.79 | 12.05 | 0.68 | 185.6 | 32.5 |

Source: Luo et al. (2003)

**Table 9.3** Permeability test of Yellow Soil

| Land use | Soil Moisture (%) | Saturated Soil Moisture (%) | Permeability Rate (mm/min) | Permeability Coefficient $K_{10°C}$ |
|---|---|---|---|---|
| Without GfG | 7.25 | 29.62 | 7.93 | 5.63 |
| With GfG | 11.47 | 38.425 | 11.49 | 9.34 |

Source: Luo et al. (2003)

water is reflected in the physical properties of the soil, in particular soil porosity. Luo et al. (2006) conducted their study in Tianquan County of Sichuan province and collected soil samples each July from 2002 to 2005. In Tianquan County, the annual mean temperature is 15.1 °C and annual precipitation is 1,735.6 mm. The physical properties of the soil are an important indicator of soil fertility. *Alnus cremastogyne* had the smallest annual non-capillary porosity but the average value of lower soil layers (20–40 cm) already reached 16 %. The largest annual non-capillary porosity was found for *Cunninghamia lanceolata* and its average value of upper soil layer is as much as 61 %. The average capillary porosity of the four vegetation types is 40.43 %. The findings suggest that all four species assessed by Luo et al. (2006) were transforming the porosity of the soil from capillary porosity to non-capillary porosity, an indicator of improving soil fertility. Similar trends were also observed in the maximum and minimum moisture capacity.

Pan et al. (2006) looked at the same issue in Zigui County (Hubei Province) which joined the GfG in 2000. Zigui County has an area of 2,472 km$^2$ and a population of 398,000. It has a subtropical continental monsoon climate with an annual average temperature of 18 °C and annual precipitation of 1,100 mm. In 1999, 126,000 hm$^2$ were affected by soil erosion (52 % of the total land area) according to statistics from the local government. The study lasted from 2000 to 2005 and investigated the impact of ten different species (Table 9.4). Apart from the physical properties of the soil, the study also assessed the water holding capacity of the soil. Pan et al. (2006) found that total and non-capillary soil water storage increased by 42.5 % and 221.4 % respectively (Table 9.4 shows the extent to which the species introduced improved the non-capillary water storage capacity of the soil). Pan et al. (2006) concluded that such an important improvement in the ability of the soil to hold water help reverse desertification.

**Table 9.4** Frequency and index of water holding capacity of soil of different conversion type of the ten most common species introduced by the GfG in Zigui County (Hubei Province)

| | Converted Area ($hm^2$) | Proportion of total converted area (%) | Total porosity | Non-capillary porosity | Total soil water storage | Non-capillary soil water storage |
|---|---|---|---|---|---|---|
| Large tangerine (Amorpha) | 2,904.5 | 22.4 | 46.3 | 2.40 | 3,700.1 | 184.3 |
| Black Locust | 1,957.8 | 15.1 | 44.8 | 2.50 | 2,688.1 | 150.1 |
| Chestnuts | 1,831.8 | 14.1 | 48.8 | 1.1 | 1,952.4 | 44.2 |
| Walnuts | 1,678.8 | 13.0 | 44.9 | 1.75 | 2,694.3 | 105.1 |
| Amorpha | 487.7 | 3.8 | 50.8 | 3.2 | 2,032.2 | 128.1 |
| Cypress(Amorpha) | 73.2 | 0.6 | 44.9 | 0.4 | 1,796.0 | 16.0 |
| Fir | 69.7 | 0.5 | 44.1 | 1.8 | 3,087.0 | 126.4 |
| Bamboo | 42.6 | 0.3 | 50.8 | 6.12 | 2,539.4 | 306.0 |
| Oak tree | 7.4 | 0.1 | 45.5 | 4.8 | 1,820.3 | 192.2 |
| Masson pine | 6.6 | 0.1 | 44.7 | 1.8 | 1,788.1 | 72.0 |
| Cropland | | | 42.3 | 1.03 | 1,691.2 | 41.2 |

Source: Pan et al. (2006)

**Table 9.5** Analysis of the influence on sediment production 2001–2002

| Year | Study Area | Annual Surface Runoff ($m^3$) | Sediment Yield (kg/100 $m^2$) | Soil Erosion Modulus (t/$km^2$/year) |
|---|---|---|---|---|
| 2001 | Pai Li Po | 14.56 | 75.53 | 755.3 |
| | Lao Ma Yuan | 14.56 | 110.2 | 1,102 |
| | Shui Jing Po | 14.56 | 96.8 | 968 |
| 2002 | Pai Li Po | 21.33 | 7.84 | 78.4 |
| | Lao Ma Yuan | 21.33 | 17.38 | 173.8 |
| | Shui Jing Po | 21.33 | 18.57 | 186.7 |

Source: Mei and Xiong (2003)

Similar positive results were found at a site in Shaanxi Province. In the Chaigou Watershed of Wuqi County, the average soil moisture and moisture-holding capacity in GfG plots after 5 years was 48 and 55 % greater, respectively, than those in non-GfG plots (Liang et al. 2006; Liu et al. 2002).

Mei and Xiong (2003) expanded such research to 10 villages in Qingzhen Town of Guizhou Province, where the GfG had converted a total area of 56.26 $km^2$ (84,390 mu) between 2000 and 2002. Mei and Xiong (2003) followed the national standards for soil and water conservation testing (SD239-87) to assess the ecological impact of the GfG. The GfG successfully conserved the soil (Table 9.5), especially when *Pennisetum hydridum* and *Silphium perfoliatum L.* were planted. In the study area, the soil erosion modulus dropped from 2,500–5,000 t/$km^2$/year in 2000 to 78.4–185.7 t/$km^2$/year in 2002, equal to a drop of 38,563.6 t of surface soil loss annually.

Yang et al. (2006) analyzed the effect of soil and water conservation on cropland that returned to forest in Wuqi county (northern Shaanxi Province) through field

**Table 9.6** Annual soil erosion moduli (t/km²/year)

| Year | Vegetation coverage (%) | Erosion modulus |
|---|---|---|
| 1997 | 19.2 | 15,280.2 |
| 2000 | 36.5 | 11,478.8 |
| 2002 | 49.6 | 8,800.7 |
| 2004 | 69.8 | 5,865.1 |

Source: Yang et al. (2006)

**Table 9.7** Bulk density of soil of different depth before and after the GfG (g/cm³)

| | Before the GfG | | | | After the GfG | | | |
|---|---|---|---|---|---|---|---|---|
| Depth | Trees | Shrubs | Grass | Average | Trees | Shrubs | Grass | Average |
| 0–20 cm | 1.45 | 1.40 | 1.42 | 1.42 | 1.30 | 1.27 | 1.38 | 1.32 |
| 20–40 cm | 1.43 | 1.45 | 1.49 | 1.46 | 1.30 | 1.32 | 1.29 | 1.30 |
| 40–60 cm | 1.38 | 1.47 | 1.48 | 1.33 | 1.21 | 1.31 | 1.27 | 1.26 |
| 60–80 cm | 1.46 | 1.48 | 1.53 | 1.49 | 1.28 | 1.33 | 1.23 | 1.28 |

Source: Yang et al. (2006)

**Table 9.8** Chemical characteristics before and after the GfG

| | pH | Organic matter (g/kg) | Total nitrogen (g/kg) | Total Potassium (g/kg) | Total Phosphorus (g/kg) |
|---|---|---|---|---|---|
| Before GfG | 8.41 | 5.9 | 0.47 | 21.3 | 1.41 |
| After GfG | 8.20 | 13.8 | 1.32 | 25.2 | 2.20 |

Source: Yang et al. (2006)

observation and experimental studies. The characteristics of the landscape in the northern Shaanxi Province Loess Hills (spurs, ridges, valleys and ravines) together with sparse vegetation coverage and concentrated precipitation, contributed to this area having the most severe soil erosion (erosion modulus 10,000–20,000 t/km²/year) in Shaanxi Province. In Wuqi county, almost 57 % of the land has a slope no less than 25° and the total area of soil erosion was 3,693 km², making it one of the counties with the greatest soil erosion. The implementation of GfG since 1998, as well as a prohibition of animal grazing, increased the vegetation cover and decreased the erosion modulus (Table 9.6). Both the physical and chemical characteristics of the soils greatly improved after implementation of the GfG, in particular the bulk density of the soil (Table 9.7), and its chemical characteristics (Table 9.8). Yang et al. (2006) concluded that returning cropland to forest and prohibiting grazing in mountainous areas are the most effective approaches to control soil erosion and water loss in the Loessy Hills region.

Yang (2005) also found significant ecological benefits of the GfG in Zhongba Village of Zijui County, Hubei Province (located in the lower section of upper Yangtze River). After the village joined the program in 2000, a total of 132.49 hm² were afforested or reforested in Zhongba Village, among which 36.37 hm² were economic forests and 84.79 hm² were ecological forests. Yang (2005) found that through the GfG, 17,258.8 $m^3$ of water and 11,158.5 $m^2$ of non-forested land were conserved. The runoff on the reforested land was 77.5 %, 85.2 % lower than the runoff on slope cropland, with a runoff coefficient between 0.0195 and 0.0296. Yang (2005) concluded that the GfG program considerably improved water conservation.

Similarly, Wang et al. (2007d) found that in Zigui County (Hubei Province), 3,085 ha of cropland (8.1 % of total cropland in Zigui County) were converted to forest in 2000, lowering soil erosion by 54,900 t a year between 2000 and 2005. Five years after the start of the GfG program, surface runoff was reduced by 75–85 % and soil erosion by 85–96 % on converted plots, compared to steeply-sloping non-GfG plots on which crops were grown.

Another study in Hunan province by Li et al. (2006) also supported earlier findings that the GfG reduces surface runoff and soil erosion. In Hunan Province, between 2000 (when the program began) and 2005, soil erosion declined by 30 %, and surface runoff dropped by approximately 20 %.

## *Impact on Desertification and Soil Erosion*

Ma and Fan (2005) argued that the GfG conversion of farmland to forestland also reduces water consumption because the land no longer needs to be irrigated. They found that 516,000 $m^3$ of water were saved in 2003 through reduced irrigation on 4,300 ha of GfG-converted land in Minqin county (Gansu Province). In that area, desertification dropped not only because of increasing tree cover, but also indirectly because tree stems and leaves can absorb air dust, reduce wind speed on the soil surface by 30–50 %, and increase air humidity 15–25 %.

Another study, however, produced different results. Zhang et al. (2011) examined the landscape-level impact of the GfG in arid environments, and found that the impact of reforestation programs on the water balance is not always positive. The research was carried out between 1998 and 2005 in a northern part of China's Shaanxi Province, where the researchers randomly selected five counties out of the total 25 that participated in the GfG program (Table 9.9). The study area has

**Table 9.9** Changes in the vegetation cover of five counties in Shaanxi Province from 1998 to 2005

| | Total vegetation cover (%) | | | | | |
|---|---|---|---|---|---|---|
| Year | Jingbian | Ansai | Baota | Yanchang | Luochuan | Average |
| 1998 | 19.5 | 22.1 | 28.5 | 21.5 | 56.9 | 29.7 |
| 1999 | 19.6 | 22.7 | 28.4 | 22.9 | 57.2 | 30.1 |
| 2000 | 21.5 | 24 | 29.7 | 24.5 | 58.2 | 31.6 |
| 2001 | 22 | 25.5 | 31.5 | 26.1 | 59.8 | 32.9 |
| 2002 | 23.7 | 27.7 | 34.8 | 28.9 | 62 | 35.4 |
| 2003 | 25.9 | 31.1 | 37.1 | 32.7 | 64.9 | 38.3 |
| 2004 | 26.4 | 33 | 39.4 | 35.9 | 66.5 | 40.3 |
| 2005 | 27.9 | 35.3 | 41 | 38.6 | 67.9 | 42.2 |

Source: Zhang et al. 2011

**Table 9.10** Soil moisture of the abandoned cropland and reforested cropland after conversion

| Length of Year | 0–1 | 1–2 | 2–3 | 3–4 | 4–5 | 5–6 | Average |
|---|---|---|---|---|---|---|---|
| Abandoned cropland percent | 16.6 | 10.48 | 10.27 | 11.42 | 10.05 | 10.04 | 11.48 |
| Reforested cropland percent | 6.14 | 5.99 | 7.49 | 7.54 | 8.13 | 8.10 | 7.23 |

Source: Zhang et al. (2011)

experienced severe soil erosion, of up to 15,000 t/km$^2$ per year. Yellow clay, with a degree of porosity of 52.1 %, is the major soil type in this area, contributing to 75 % of the total. The total vegetation cover in areas covered by the GfG increased from 29.7 % in 1998 to 42.2 % in 2005. However, the survival rate of the forests averaged only 49 % in the seventh year. Further, Zhang et al. (2011) found that abandoned cropland may retain water better than trees planted through reforestation programs (Table 9.10). They concluded that abandoned lands and natural vegetation recovery can retain the water in the soil at the lowest cost. Reforestation can achieve similar results only when the species of plants are carefully selected and the reforested lands are monitored and tested for a considerable period of time. The study concluded that afforestation may remain a valuable tool but should be limited to the planting of native or other species that will not exacerbate soil water shortages. These may include stable communities of natural desert steppes, or dwarf shrubs that maximise water-use efficiency, and possibly even lichen species in more severely degraded environments (Zhang et al. 2011).

In yet another study, Wang et al. (2007c) argued that the conversion of cropland into forestland has been overly emphasized in arid areas, and that it would be better to grow native plants of grass or scrubs. Through a survey of 208 counties in desertification-affected northern China, they found that few areas have water conditions that are suitable for planting trees: 88.3 % of the study area has an arid or semi-arid climate with an annual precipitation of less than 400 mm, which is suitable for grass or scrub growth but unsuitable for forest growth (Wang et al. 2007c). Although some successful cases have been reported, the overall survival rates of planted trees are very low, about 30 % (Jiao 2005), and dropping to only 10 % in some areas (Shen et al. 2003). While Uchida et al. (2005) considered water deficit problems as the most important cause for the low survival rate and slow growth rate of the vegetation, Wang et al. (2007c) argued that the inappropriate choice of tree species, careless planting and inadequate management were equally important. Wang et al. (2007c) argued that large-scale afforestation may also cause potential environmental problems, such as increasing evapotranspiration and intensification of soil erosion, desertification and sandstorms, all of which result in great waste of manpower and money. Compared with planting trees, planting drought-tolerant grasses and scrubs requires much lower investment and are favorable to most of northern China (Wang et al. 2007c).

The findings of Wang et al. (2007c) were further supported by a survey conducted by Zhang et al. (2011). Zhang focused on five randomly selected counties (Jingbian, Ansai, Baota, Yanchang, Luochuan) from the 25 counties in northern Shaanxi

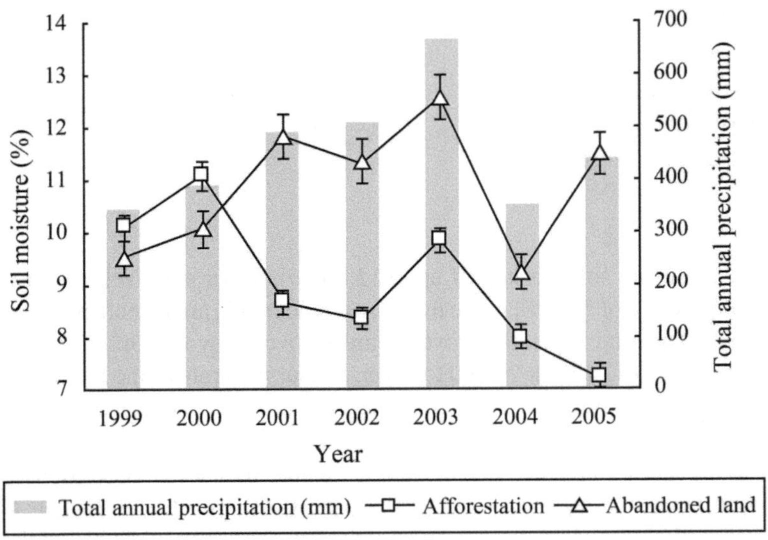


**Fig. 9.1** Changes in total soil moisture (%) at a depth of 6 m during the growing season, 1999 to 2005 (Source: Zhang et al. (2011))

Province that are covered by the GfG. Historically, severe soil erosion has occurred in these areas, at an average rate of approximately 15,000 t per km$^2$ per year. From 1998 to 2005, total annual precipitation averaged 461.7 mm, ranging from 366 mm in Jingbian County to 609.4 mm in Luochuan County (Cao et al. 2009a; Zhang et al. 2011). During the same period, the potential evapotranspiration in the study area averaged 793.7 mm per year, well above average precipitation.

Zhang et al.'s results (2011) suggest that the policies of prohibiting cultivation and grazing in steep terrain were significantly more effective than the afforestation policy. This finding offers a valuable strategy for environmental restoration in similar remote rural regions, both in China and around the world (Zhang et al. 2011). Figure 9.1 summarizes the changes pertaining to soil moisture from 1999 to 2005 (afforestation in the region started in 1998). A linear regression of the total annual precipitation against the soil moisture shows a strong and positive correlation in the abandoned land plots ($R=0.91$, $p<0.01$), but a very weak and nonsignificant correlation in the afforestation plots ($R=0.01$, $p>0.05$) (Fig. 9.1). Although correlation does not imply a causal relationship, the findings could be explained as follows: when grassland is restored using unsuitable tree species, there may be insufficient precipitation to permit a balance between the available soil moisture and the vegetation cover, leading to a risk of declining soil moisture (Zhang et al. 2011). The researchers concluded that large-scale afforestation in this vulnerable arid and semi-arid region could increase the severity of water shortages, decrease vegetation cover in afforestation plots, and adversely affect the number of species present. The exclusion of livestock from overgrazed areas and the elimination of cultivation in marginal areas had the biggest positive effects on the restoration of vegetation cover, whereas tree planting had a strong negative effect in vulnerable areas (Zhang et al. 2011).

## Carbon Sequestration

While studies on the consequences of the GfG on soil recovery and water conservation require extensive fieldwork and therefore are predominantly site specific, studies on carbon sequestration are often done using satellite images and existing international, national and provincial datasets, and do not require fieldwork. For this reason they are often carried out on a much larger scale. In the following section we review two such studies, one done on Yunnan province, the other on the whole of China.

In Yunnan Province, after a pilot phase in 2000–2001, the GfG was implemented in 126 counties, which corresponds to most of the province. Over 95 % of forests established through the GfG in Yunnan Province were ecological forests. The Yunnan Provincial Forestry Department (YPFD) estimated the carbon sequestered through the GfG in the province by using data on the area of tree species planted during 2000–2007 (Chen et al. 2009). The department developed four scenarios for GfG area stands to be planted annually between 2008 and 2010, and options for harvesting the trees. According to technical regulations for ecological forests, these trees may not be cut until they are mature. so, the basic assumption was that planted forests are not harvested until mature. The carbon sequestration potential of these converted forests is expressed as the carbon stock changes in the tree biomass and soil organic matter. The GfG lands, are largely degraded croplands or barren lands that generally have a low initial Soil Organic Carbon (SOC) stock. The YPFD developed empirical growth curves for different tree species, based on data from the National Forestry Inventory on the growth of existing forests in Yunnan Province. These growth curves were then used to estimate the carbon stocks in the tree biomass pools, using basic wood density, biomass expansion factors and carbon fractions. Empirical factors were also introduced to estimate the stock change in SOC under the GfG.

The YPFD found that the GfG would contribute significantly to carbon sequestration in Yunnan Province (Table 9.11), whether the area planted for each species is estimated using Scenario A, which uses the planned goal of reforestation by the GfG, or Scenario B, which uses the average annual area reforested from 2005 to 2007. Scenario A implies that the reforestation (and restoration of the original vegetation) goals of the governmental will be fulfilled, and the planting area (including the area converted to grassland) of the GfG will be up to 1.238 Mha. In this case, the carbon sequestered will be up to 54.128–56.621 TgC by the year 2050 (Chen et al. 2009). Under scenario B, the area planted by the GfG will be up to 1.139 Mha and the carbon sequestered will be up to 49.918–52.083 TgC by the year 2050 (Chen et al. 2009). The carbon sequestered by the seven major tree species accounts for 43.27–50.56 % of the total carbon sequestered through GfG-led land use/land cover changes. By 2050, the total carbon sequestered by the vegetation introduced through the GfG is expected to be up to 10.82–12.27 % of the carbon stocks of

**Table 9.11** Area planted and carbon sequestration of the three most common tree species/species group under the GfG Program in Yunnan Province

| Tree Species | Planted area 2000–2007 (ha) | Carbon Sequestration Potential (TgC) | | | | |
|---|---|---|---|---|---|---|
| | | 2010 | 2020 | 2030 | 2040 | 2050 |
| *Pinus armandii* | 77,369.8 | 0.436–0.442 | 3.505–3.624 | 5.399–5.829 | 5.965–6.468 | 6.175–6.715 |
| *Eucalyptus* spp. | 74,116.1 | 1.172–1.187 | 2.822–3.024 | 3.612–3.905 | 4.127–4.472 | 4.374–4.755 |
| *Pinus yunnanensis* | 65,193.2 | 1.111–1.121 | 2.977–3.166 | 4.151–4.472 | 4.879–5.279 | 5.205–5.655 |

Source: Chen et al. (2009)

forest ecosystems in Yunnan province in the 1990s (Chen et al. 2009). Table 9.11 displays the data for the three most common species.

Using official statistics from the program, Ostwald et al. (2011) estimated the nation-wide amount of carbon that has been sequestered by the GfG. They collected information from forestry statistics at the national and province level and from the scientific literature on the locations of plantations, the physical characteristics of the locations, the species planted, the rate of increment per year, and survival rates. To estimate the carbon sequestration performance of the GfG, they established a baseline of what would plausibly have happened in the absence of the program. Due to the targeted soils' degraded character with high erosion and unsustainable agriculture, the soils were assumed to contain no carbon when the program was initially implemented. Carbon sequestration was then calculated according to three different approaches based on (1) net primary production, (2) figures from IPCC's greenhouse gas inventory guidelines, and (3) mean annual increment. The carbon pools included in the calculation were above and below ground biomass, with the latter at a ratio of 0.26 to the former (Ostwald et al. 2011).

The calculation, done in 2009, revealed that conversion of cropland and barren land over the first ten years generated carbon sequestration ranging from 222 to 468 million tons of carbon (MtC), with the IPCC approach yielding the highest estimate (312 MtC). The other two approaches showed similar results (around 250 MtC). The median of 246 MtC corresponds to 14 % of the carbon emitted in the year 2009. This would mean an annual sequestration range from 22–47 MtC per year, with a median of 25 MtC. If taken on a hectare basis, a carbon content of 11–23 tC per hectare indicates low productivity. Sichuan has the largest amount of carbon sequestered through the GfG-induced land cover/land use change, with 31.7 million tons of carbon, while Tibet has the lowest, with 209,000 t of carbon (Fig. 9.2). Figure 9.2 shows that in most provinces a similar amount of carbon was sequestered through the transformation of barren land and cropland, though slightly more had been sequestered through the transformation of barren land. Ostwald et al. (2011) estimated that nationwide, 53.6 % of the carbon sequestered through the GfG between 1999 and 2008 was on land that had previously been barren, while 46.4 % was on former cropland. Only Xinjiang, Qinghai, Shaanxi, Sichuan and Jilin have larger carbon sequestration through cropland conversion than through barren land conversion (Ostwald et al. 2011).

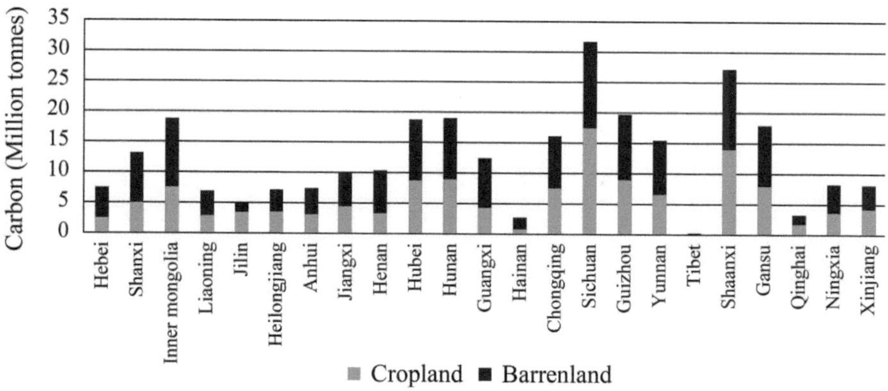


**Fig. 9.2** Average amount of carbon sequestrated by conversion of cropland (46.4 %) and barren land (53.6 %) under the GfG 1999–2008 (Source: Ostwald et al. 2011)

## Conclusions

This chapter has reviewed some of the problems with the GfG, focusing on the conservation and improvement of soil conditions and fertility, conservation of water resources, and carbon sequestration. We have argued that the ecological consequences of the GfG have generally been positive, especially in relation to soil fertility and the improvement of soil conditions. The GfG has also been positive in terms of improving water balance and reducing siltation. However, there is considerable controversy around the consequences of the GfG in arid areas, centered around the choice of planting trees. In these areas, the trees' survival rate is low, evapotranspiration increases and may be higher than the precipitation, and the water table is further reduced. Shrubs and native vegetation may be better choices in these areas. On the other hand, the GfG has contributed considerably to carbon sequestration, especially through planting ecological trees, which may not be cut until mature. While researchers agree that the GfG overall has had a positive ecological impact, broader, nation-wide conclusions are difficult to draw because of the absence of nation-wide studies. Except for carbon sequestration, studies are limited to small case studies, but the extent to which findings may be extrapolated to the rest of the country is questionable, given the diversity of ecological, climatic, and socio-economic conditions in China.

# Chapter 10
# Impact on Grain Output and Price

**Abstract** This chapter looks at the impact of the Grain for Green on the grain output and price. Between June 2002 and June 2004, grain prices increased, which was blamed on the reduction of farmland set aside by the Grain for Green. As a consequence, the Grain for Green was scaled down after 2004. However, most researchers agree that the Grain for Green was not responsible for the increasing price of grain, because most of the land converted by the Grain for Green was not very fertile. Furthermore, farmers could concentrate their efforts on their more fertile parcels, with the result that grain production overall fell only very marginally in Grain for Green-converted areas. Instead, researchers argued that the drop in grain output, and subsequent increase in grain prices, was more likely due to the loss of farmland caused by urbanization in the eastern provinces (where the Grain for Green was not implemented), as well as the shift from grain production to the production of other crops.

**Keywords** Price of grain • Grain output • Food security • Land productivity • Land fertility

## Introduction

In 2003 production of grain in China declined and its price increased. Many blamed this on the Grain for Green program, since the GfG had converted millions of hectares of farmland to forest during the previous years, culminating with the conversion of 3.4 million hectares of farmland in 2003 alone (Table 3.3). The blame was later found to be unjustified, but the decline in grain production was understandably of great concern in China, which, even in good years, is able to produce only a small surplus of food. This could be a significant problem in a country with nearly 1.3 billion people in 2003 and a population that was continuing to grow.

In this chapter we review the literature that discusses the impact of the GfG on the amount of farmland, the quantity of grain produced, and the price of grain. While most researchers agree that the GfG program was not the primary contributor to the drop in China's overall grain output, it did appear to have an impact in some regions with limited land, where it led to localized shortages, aggravated by insufficient infrastructure to transport the grain deficit to the local population.



C.O. Delang, Z. Yuan, *China's Grain for Green Program*,
DOI 10.1007/978-3-319-11505-4_10

Overall, the analysis of factors contributing to a loss of grain output has highlighted the tenuous relationship between population growth and food requirements. The analysis suggests that, if managed properly, an environmental program that focuses on land recovery can be beneficial to the population. The program, however, should be part of an overall planning process that includes urban planning, population control, and investment in infrastructure.

## Grain Output, Price, and Farmland Area

From 1965 to 1999, when the GfG was introduced, the total amount of productive farmland in China remained fairly stable, with land that had lost fertility being replaced by similar amounts of land cleared from forests. In 1999, China had 1.3 % more farmland producing grain and soybeans than in 1965 (Fig. 10.1). After 1999, the year the GfG began implementation, however, there was a sharp drop. Sown area dropped from 97.7 million hectares in 1999 to 84.7 million hectares in 2003, a drop of 13.3 %. In fact, in 2003 the area sown with grain and soybeans by China's farmers fell below the 85 million hectare level for the first time since the 1950s. The reduction in productive farmland can be compared to the farmland set aside through the GfG: in Chap. 3 (Table 3.3) we reported that 13.1 million hectares were reforested by the GfG from 1999 to 2003, at least 5.8 million hectares of which were converted cropland.

China's grain production trends parallel those of sown areas and reinforced the concerns about the effect of the GfG. After trending up from the 1960s to the 1990s, grain production in China dropped from 392.29 million tons in 1998 to 322.96 million

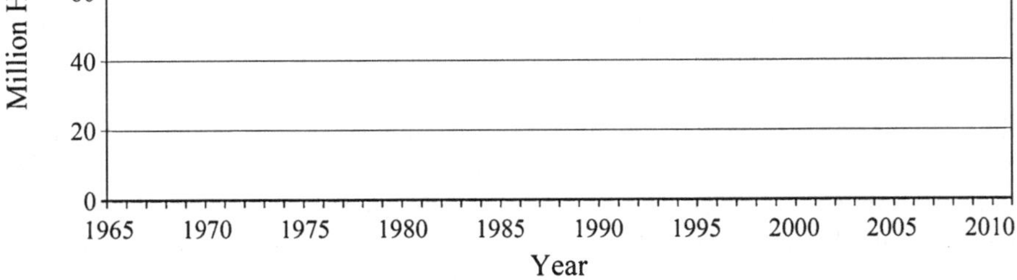


**Fig. 10.1** Grain and soybean area harvested in China, 1965–2011 (Source: Earth Policy Institute 2013a)

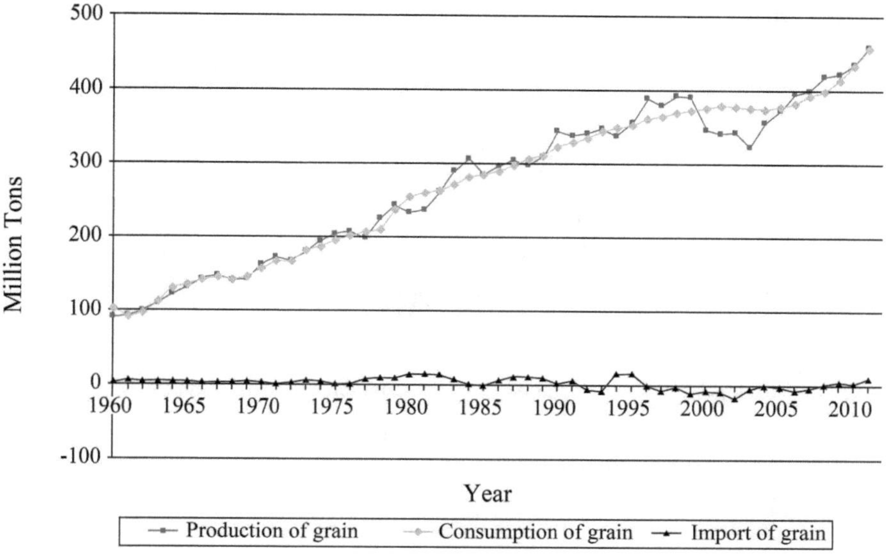


**Fig. 10.2** Chinese production, consumption and imports of grain (1960–2011) (Source: Earth Policy Institute 2013b)

tons in 2003, a drop of 17 % over 5 years (Fig. 10.2).[1] From 2000 to 2003 grain production was well below the consumption level. However, because of population growth, grain per capita decreased even more, from 285 to 227 kg over the same period, a drop of 20 % over the 5 year period.[2] Only in 2006 did production return to 1999 levels. Not surprisingly, it was assumed that the shrinking croplands had jeopardized Chinese food security.

The issue is also of concern to other countries. China is the largest grain consumer in the world. A drop in production in China would eventually need to be compensated by imports from other countries, with consequent impact on the world price of grain. Thus, the grain issue has a direct bearing on China's economic development and national security, as well as international grain security (State Administration of Grain 2002).

The drop in the amount of grain produced was accompanied by an increase in the price of food. From January 2003 food prices steadily increased, and by June 2004 food prices were up 15 % annually (Fig. 10.3). China's rate of food inflation began to drop in June 2004 but during 2003 and 2004 food inflation was blamed on the GfG.

The drop in grain production and the increase in food prices were blamed on the GfG. (Ministry of Land and Resources 2004). Since the GfG had converted a

[1]At the same time, from 1996 to 2007, China was a net exporter of grain, which culminated with 17.1 million tons of grain exported in 2002 (Earth Policy Institute, 2013). It seems that, although production dropped from 2000, the situation was not considered too alarming, although, as Tao et al. (2004) say, the national food stocks gradually shrank.

[2]Data from the Earth Policy Institute (http://www.earth-policy.org/data_center/C24). Different sources present different data, but the proportions are similar.

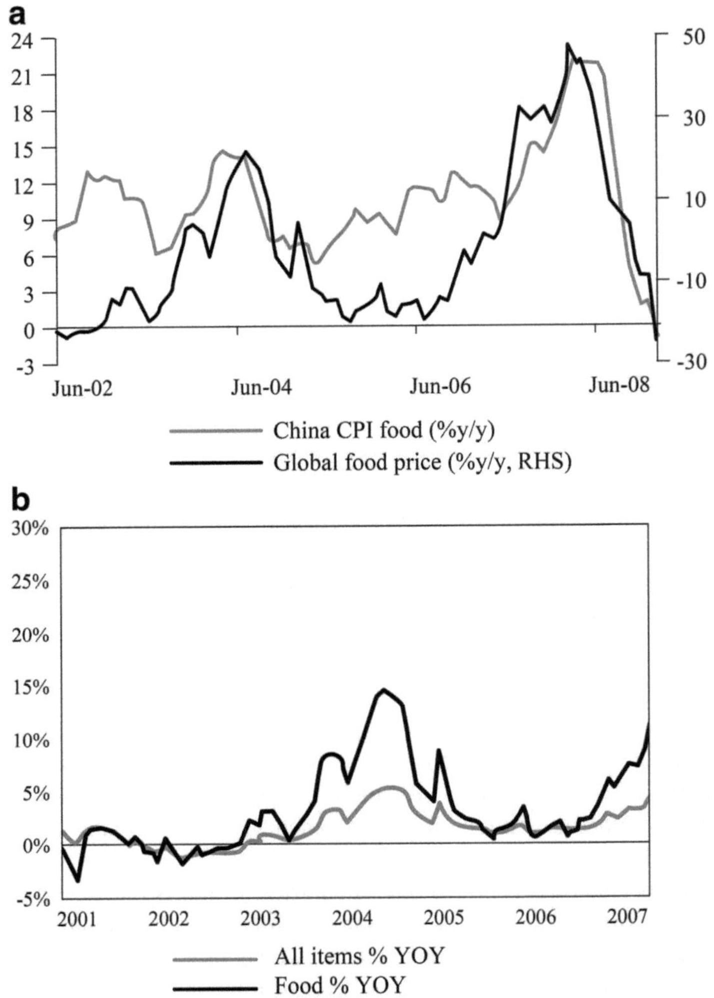


**Fig. 10.3** Changes in food prices (2002–2007) (Source: Zhu 2008)

considerable amount of farmland, it is not surprising that many people blamed the program. The possible relationship between the conversion of farmland by the GfG and the drop in grain harvest levels was confirmed in research by Dong et al. (2010), who overlaid GfG remote sensing maps and agricultural suitability maps to determine whether the converted areas were in the low suitability level. From the maps, Dong et al. (2010) claimed that GfG was one of the main reasons for cropland loss.

These findings at the national level were confirmed by Zhou et al. (2007) at a county level in Liping County (Guizhou province). Zhou et al. (2007) found that over a period of 14 years, between 1989 and 2003, the production of grain overall increased by close to 25 % (Fig. 10.4). However, this was matched by population growth, and the grain output per capita in 2002 was almost the same as that of 1989, only 281 kg.

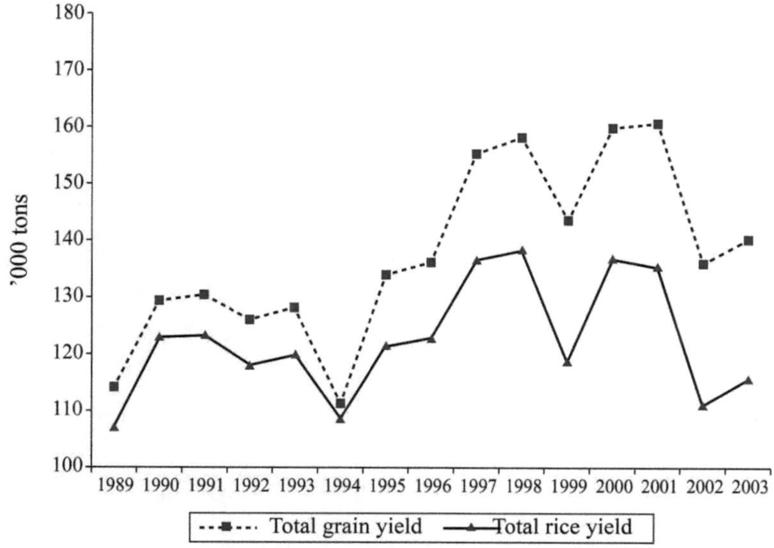


**Fig. 10.4** Change of grain and rice yields in Liping (Source: Zhou et al. 2007)

On the other hand, while the production of rice fluctuated considerably during the period under consideration, by 2003 it was at the same level as 1989 (Fig. 10.4), which means that per capita output had dropped considerably from 1989 to 2003. According to Zhou et al. (2007) after the implementation of the GfG program in 2000, the food production capacity decreased in Liping County, because of the removal of cultivated land from agriculture. Zhou et al. (2007) concluded that the implementation of the GfG project in 2000 resulted in a decline in food sufficiency in Liping County.

In China, the belief that land conservation contributed in a major way to the deterioration of the nation's food security was so strong, that the leadership severely curtailed the expansion of the program in 2004. However, many researchers disputed the relationship between the GfG and the drop in the grain harvest. For example, according to Xu and Cao (2002), in most of the regions where the program was operating, the productivity of the land converted since 1999 was lower than that of the non-converted land. In addition, the productivity on the plots that remained under production seemed to increase after the implementation of the GfG. This means that the GfG could not be blamed for the drop in the amount of grain produced. Indeed, many argue that the decrease in cultivated land was not the main cause of the reduction in grain output, and that rather than the GfG, other factors were the real culprits.

First, according to Li Zibin, vice minister of the National Development and Reform Commission, there has been a trend toward using more farmland for crops other than grain, which contributed to the decline in the grain harvest (China Daily 2004).

Second, China experienced accelerating urbanization and industrialization (Cai et al. 2002) that inevitably resulted in a transformation of large amounts of cultivated land into new industrial and urban areas. The competition over farmland by these more profitable uses of the land contributed more to the reduction in grain output than did the GfG. Coastal provinces like Jiangsu and Guangdong, which did

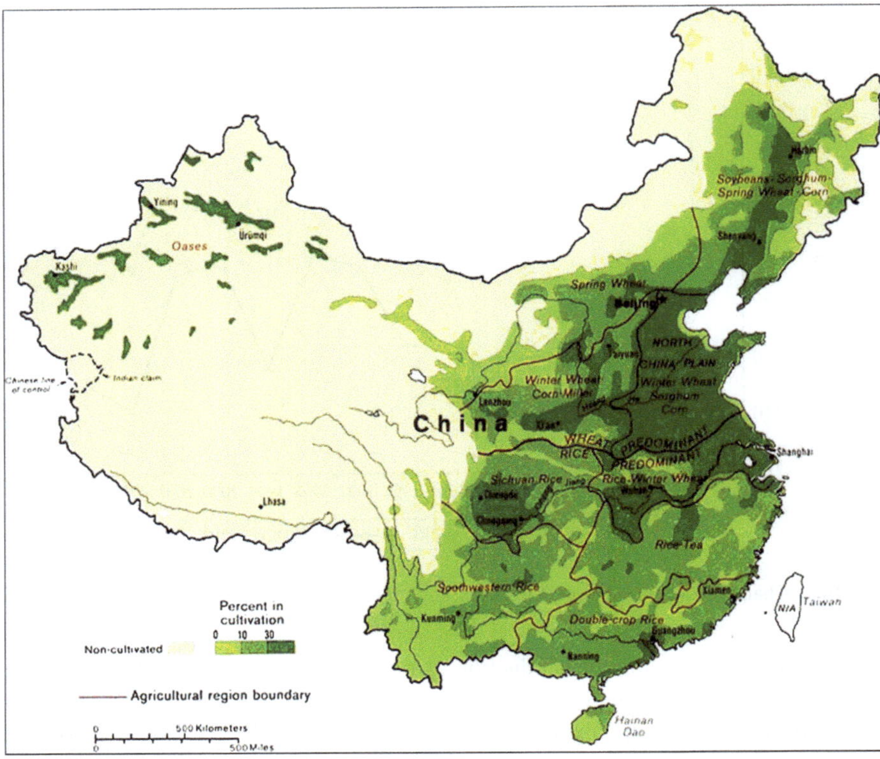


**Fig. 10.5** Agricultural regions of China (Source: U.S. Central Intelligence Agency 1986)

not participate in the GfG, saw grain outputs drop by nearly 30 million tons in 2003 from 1998 levels (China Daily 2004). With the development of the economy, we would expect cultivated land in eastern China, where the land is most productive, to further decrease in the future, (Fig. 10.5) (Hong and Li 2000).

Third, according to Tao et al. (2004), experts believe that the drop in production had to do with depressed grain prices, weak public investment in agriculture, and the unsuccessful attempt to reform the nation's grain marketing system in the late 1990s and early 2000s. According to Tao et al. (2004), these factors are more likely than the GfG to have made prices for food and grain increase from June 2003 to June 2004.

To these factors has to be added the fact that the farmland converted to the original land cover was predominantly unproductive or marginal. Xu et al. (2006b) looked at the production impacts of the implementation of China's GfG program since the pilot program began in 1999. They made the point that the GfG mainly targeted steeply sloped land in poor, remote, and mountainous regions where productivity was almost certainly much lower than the areas in which land was not retired. Indeed, yields on the GfG plots were on average only about 30 % of those of the non-GfG plots. GIS databases further consolidate the lower productivity

**Table 10.1** Simulated impact of GfG policy on wholesale prices of agricultural commodities

| Commodity | Period | Actual price in 2003 (Yuan/t) | Change in percentage (percent) | | |
|---|---|---|---|---|---|
| | | | Change over period | Change due to | |
| | | | | GfG policy | Other |
| Rice | 1999 | 1,659 | 1.57 | 0.30 | 1.27 |
| | 2003 | 1,685 | | | |
| Wheat | 1999 | 1,458 | −7.06 | 1.85 | −8.92 |
| | 2003 | 1,355 | | | |
| Maize | 1999 | 1,117 | 8.86 | 1.70 | 7.16 |
| | 2003 | 1,216 | | | |
| Other coarse grains | 1999 | 1,375 | −6.91 | 2.25 | −9.16 |
| | 2003 | 1,280 | | | |

Source: Xu et al. (2006b)

potential of GfG counties compared to those of non-participating counties. Even from an economic perspective, it seems that the losses from conversion of agricultural land were relatively small. Uchida et al. (2005) reviewed the cases of Ningxia and Guizhou, and concluded that, in spite of the households' setting aside over half their land, the per capital net incomes from grain decreased by only 25 % and 11 % respectively.

Xu et al. (2006b) concluded that, at most, the direct loss of production from the GfG was only about 9.6 % of the total drop in output. They also adapted a quantitative method based on CCAP's Agricultural Policy Simulation and Projection Model (CAPSiM) to model the full effects of the GfG program on China's grain production. One drawback of CAPSiM is that it is a partial equilibrium model in the sense that it looks only at the agricultural sector and does not include factor markets. Nevertheless, it is the first and most comprehensive model to examine the effects of policies on China's food demand, supply, and trade (Xu et al. 2006b). The simulation analysis shows that grain prices do rise due to the GfG, but by almost any point of view the contribution of the GfG to price increase is small compared the total change in price (Table 10.1). Xu et al. (2006b) concluded that the GfG had only a negligible impact on the price increases of 2003–2004; it did not lead to a reduction in national food security, and it did lead to a negligible rise in grain imports, accounting for less than 0.05 % of production or consumption.

The lower productivity of set-aside land is not always given. As we discussed in Chap. 5, in some cases highly productive land has also been set aside. For example, Wang et al. (2007c) claimed that farmers not only converted steep cropland, but also converted some fertile cropland into forests to obtain additional subsidies. However, according to Li Zibin, although the land reforested previously provided 6.5 million tons of grain, production efficiencies enabled the remaining farmland to increase output by 5 million tons. This resulted in a net reduction of only 1.5 million tons (China Daily 2004).

**Fig. 10.6** Western China and seven agricultural ecological zones (Source: Feng et al. 2005b)

Feng et al. (2005b) simulated the impact of the GfG program on China's grain supply in the upper reaches of the Yangtze River and the upper and middle reaches of the Yellow River. The upper reaches of the two largest rivers in China, the Yangtze and the Yellow, are critical to the middle and lower reaches of the two rivers, which cover almost half of China's territory (Feng et al. 2005b). The survey covered 2,843 counties, 43,000 towns, 740,000 villages, 25,000 farms, and 400,000 administrative units. Systematic, county-level data on the types, area, and locations of land use were added to data about cultivated sloping land from the 1996 national land survey, conducted between 1984 and 1996 using the most recent aerial photos, Landsat images, and available maps. All the cultivated land was classified into five types by degree of slope: <2°, 2°–6°, 6°–15°, 15°–25°, and >25°. This survey provided the most systematic, comprehensive, and coherent quantification measurement of China's land, given the technology available (Feng et al. 2005b).

Feng et al.'s (2005b) simulation divided the whole region into seven agro-ecological zones with similar characteristics, including climate, geographic condition, and grain production situation (Chen 2001) (Fig. 10.6). Within each zone, the productivity per hectare was estimated considering the slope of the land. Altogether, 118 rural households were interviewed (Feng and Zhang 2002). The relationship between the slope and grain production by zone from the survey is presented in Table 10.2.

To estimate the loss of grain production, Feng et al. (2005b) use data on the loss of land and the index of grain productivity of the converted land presented in

**Table 10.2** Grain production distinguished by different land slopes and averaged per region

| Region | Number of households surveyed | <2° kg/ha | 2–6° kg/ha | 6–15° kg/ha | 15–25° kg/ha | >25° kg/ha | Terrace kg/ha |
|---|---|---|---|---|---|---|---|
| Loess Plateau | 50 | 4,500 | 4,125 | 2,850 | 1,313 | 1,125 | 3,375 |
| Inner Mongolia Plateau | 5 | 5,250 | 4,350 | 3,375 | 1,200 | 900 | 3,150 |
| Qinghai-Tibet Plateau | 5 | 5,625 | 4,500 | 3,375 | 2,400 | 1,125 | 4,125 |
| Sichuan Basin | 20 | 9,000 | 7,500 | 6,000 | 2,775 | 1,875 | 6,000 |
| Yungui Plateau | 10 | 6,000 | 4,500 | 3,750 | 2,250 | 1,500 | 3,750 |
| Hengduan Mountains | 8 | 4,875 | 4,125 | 3,000 | 2,250 | 1,500 | 3,000 |
| Northwest dry region | 20 | 7,500 | 5,625 | 4,500 | 1,500 | 750 | 3,750 |

Source: Feng et al. (2005b)

Table 10.2. The total loss in grain production is the sum of production loss of all converted farmland. Feng et al. (2005b) developed three scenarios at the county level: (A) retaining 0.133 ha of agricultural land per capita; (B) retaining 0.1 ha per capita; and (C) retaining 0.067 ha per capita. In each scenario, the converted slope land would be shifted to forests and grasslands. The 1996 grain production level was used as the base level. Table 10.3 shows estimated grain production loss for each province with each of these three scenarios.

Feng et al. (2005b) estimated the loss in grain production due to the GfG to range from 2 to 3 % nation-wide. However, the impact changes drastically among regions, and at the local level, impact could be significant. For instance, the impact in Sichuan Basin is much larger than in the Loess Plateau. At the provincial level, in Guizhou, Yunnan, Shaanxi, Gansu, and Qinghai, grain losses are very high in scenarios B and C. If scenario B, and in particular C were implemented, per capita grain production of all these provinces would decrease considerably. The policy would also have great impact in some counties. Overall, in western China, there are 22 counties with grain loss expected to reach 50 % in all three scenarios (Feng et al. 2005b).

Because of the potential localized issues, Feng et al. (2005b) recommended adapting the policy to local conditions in order to address the risk of food insecurity in particular areas and reduce the amount of land to be set aside in the more food-insecure areas. They also recommended integrating the GfG with other policy and economic reforms, suggesting that the best long-term strategy was to encourage population migration from environmentally sensitive areas to areas with higher population capacity. Finally, they suggested that converting cultivated land to forest should be coordinated with improvements in the productivity of the remaining cultivated land, infrastructure development, and other resource management. A well-planned transportation system can eliminate or reduce the variation in crop losses among different regions. Yet, food redistribution between counties is costly, faces

**Table 10.3** Estimated grain production loss associated with shifting land use on province level

| Province | Scenarios | Converted land (1,000 ha) | Of total (%) | Estimated loss (1,000 t) | Of total (%) |
|---|---|---|---|---|---|
| Chongqing | A | 340 | 13.21 | 934 | 7.97 |
| | B | 340 | 13.21 | 934 | 7.97 |
| | C | 382 | 14.84 | 1,089 | 9.29 |
| Sichuan | A | 539 | 7.93 | 1,259 | 3.61 |
| | B | 539 | 7.93 | 1,259 | 3.61 |
| | C | 833 | 12.27 | 2,317 | 6.65 |
| Guizhou | A | 875 | 17.28 | 930 | 9.19 |
| | B | 953 | 18.81 | 1,066 | 10.53 |
| | C | 2,132 | 42.10 | 3,134 | 30.95 |
| Yunnan | A | 778 | 11.85 | 760 | 6.10 |
| | B | 1,879 | 28.61 | 2,395 | 19.22 |
| | C | 2,700 | 41.11 | 3,614 | 29.00 |
| Tibet | A | 6 | 1.74 | 5 | 0.59 |
| | B | 26 | 7.43 | 28 | 3.56 |
| | C | 26 | 7.43 | 28 | 3.56 |
| Shaanxi | A | 1,238 | 21.49 | 1,229 | 10.10 |
| | B | 1,733 | 30.09 | 1,840 | 15.12 |
| | C | 2,202 | 38.24 | 2,420 | 19.88 |
| Gansu | A | 1,435 | 27.77 | 1,205 | 14.68 |
| | B | 1,435 | 27.77 | 1,205 | 14.68 |
| | C | 1,435 | 27.77 | 1,205 | 14.68 |
| Qinghai | A | 6 | 0.83 | 5 | 0.40 |
| | B | 124 | 18.24 | 195 | 15.75 |
| | C | 130 | 19.19 | 205 | 16.59 |
| Ningxia | A | 159 | 12.63 | 112 | 4.36 |
| | B | 159 | 12.63 | 112 | 4.36 |
| | C | 159 | 12.63 | 112 | 4.36 |
| Xinjiang | A | 1 | 0.03 | 1 | 0.01 |
| | B | 1 | 0.03 | 1 | 0.01 |
| | C | 1 | 0.03 | 1 | 0.01 |
| **Total** | **A** | **5,377** | **14.13** | **6,440** | **6.30** |
| | **B** | **7,189** | **18.89** | **9,304** | **8.84** |
| | **C** | **10,001** | **26.28** | **14,125** | **13.83** |

Source: Feng et al. (2005b)

bureaucratic inefficiency, and there is a shortage of transportation infrastructure. Overall, however, Feng et al. (2005b) concluded that the proposed policy should not have a major impact on China's future grain supply and the world grain market.

Feng et al.'s (2005b) model does not include the impact that increased mechanization and use of chemical fertilizers and pesticides would make on agricultural output, in particular the fact that their use per unit of land will likely increase as total land decreases. In addition, the ecological and environmental improvements on the set-aside land are likely to have some impact on the remaining cultivated land, for example, on the availability of water. Similarly to other researchers reviewed above, they conclude that the GfG program has not caused a drop in grain production that is proportionate to the retired cropland, and that this is likely the result of two factors: first, the retired cropland tends to be marginal, low-yield plots, and second there is potential gain in agricultural productivity on the land that was not set aside.

These issues have also been addressed by Shi and Wang (2011) in a study of Mizhi County, Shaanxi Province in northwest China, where total grain yields reached 80,194 t in 2008, a 72 % increase over 1998 (Anonymous 1998, 2002).[3] Shi and Wang's (2011) discussions with local officials and farmers revealed the reasons for increased agricultural output despite the conversion of agricultural lands to forestland or grassland:

1. The farmlands that are converted to forestlands and have a slope greater than 25° are largely marginal and among the least productive.
2. The environmental improvements reduce the risk of natural disasters, and consequently increase the grain yields.

Yao and Li (2010) used a slightly different approach and examined the agricultural productivity change induced by the GfG using the Malmquist index method and household data collected from Wuqi county of Shaanxi Province. They found that during the period of 1998 to 2004, the total factor productivity (TFP) grew by 15.8 %. While numerous households suffered a TFP decline, the majority of them experienced a large gain. By deconstructing the TFP, they further show that its increase is due exclusively to the improvement of technical efficiency rather than to technological change. These findings are validated by estimating the TFP change with county-level aggregate data (Yao and Li 2010). Because of the tremendous cropland reduction and production mode shift caused by implementing the GfG, the TFP declined substantially during the first 3 years of the program. However, due to continued improvement of technical efficiency, its growth accelerated later (Yao and Li 2010).

Altogether, their evidence consistently suggests that implementing the GfG has contributed to the agricultural TFP growth in the longer term and that the efficiency improvement has resulted mainly from the increased public expenditures for extension services and diffusion of technical knowledge. Wuqi's experience proves that it is possible to achieve environmental conservation and increase productivity simultaneously, even when facing cropland reduction and changes in production modes (Yao and Li 2010).

[3] In spite of these general improvements, Shi and Wang (2011) found that because of its poor ecological environment, the Loess Plateau as a whole has not been able to increase its food production sufficiently to keep up with its rapidly increasing populations.

## Food Security

In 2004 agricultural tax reductions and other incentives contributed to a recovery of the agricultural output to 469.5 million tons of grain. In 2005, the total planting area reached 104.3 million hectares, and the aggregate production exceeded 484 million tons of grain. Output thereafter continued to rise steadily, reaching 528.5 million tons in 2008. The total output of grain in the summer of 2009 was 123.35 million tons, an increase of 2.60 million tons, up 2.2 % from the summer of 2008. Of this total, 2.5 million tons of grain, which accounted for more than 96 % of the total increase, resulted from expanding acreage (Fig. 10.1) (National Bureau of Statistics 2009).

The critical question emerges regarding the extent to which food security is guaranteed by the amount of cultivated land in China. Yu and Lu (2006) tried to answer this question by taking into account the driving factors behind farmland loss, including the expansion of built-up areas, natural disasters, the ecological conversion of marginal cropland, and farmland gain by land consolidation and reclamation. They assumed that (1) basic food demand should be completely self-supplied in China and (2) population growth is a key driving force of cultivated land loss along with urban and rural area expansion (urban and rural built-up area per capita was used to estimate future developments). An extrapolation trend line based on consistent data of long time series was used to predict variables' change, and coefficient of determination ($R^2$) was calculated to indicate the accuracy of the analysis (Yu and Lu 2006). Yu and Lu (2006) used data from national statistic yearbooks from 1989 to 2003 (National Bureau of Statistics of China 1990–2004). The base year was 2003 and the target year was 2030.

Yu and Lu (2006) addressed the same question of food security up to 2030, and found that China's food supply can be maintained only at a low to middle level of 370–410 kg per capita. That is, China has enough land to meet its primary food demand. However, it cannot reach the safe target of 500 kg per capita if population growth and the expansion of built-up land are not strictly controlled, and if there is no breakthrough in controlled crop breeding, or no significant improvement in irrigation works (Yu and Lu 2006).

Long and Zou (2010) looked at the same issue of food security using an approach they called the Farmland Use Level (FUL). To do so, they used data from China's statistical yearbooks and all provincial statistical yearbooks, divided the country into six regions according to geographical locations and similar physical conditions, and estimated the FUL at the provincial level (Fig. 10.7). Grain productivity was determined not only by the physical conditions of the land, but also by human investment, such as man-made materials, technology, and capital. This is becoming more important, as new methods, technologies, and crops are developed to increase the yield per unit of land. The complex change of farmland use on a large scale makes it difficult to use traditional methods to classify and assess the quality of the farmland over a large area. Long and Zou (2010) used the FUL to replace the role of farmland quality for measuring grain production. To assess the FUL, a three-layer assessment indicator system was adopted, to include investment density,

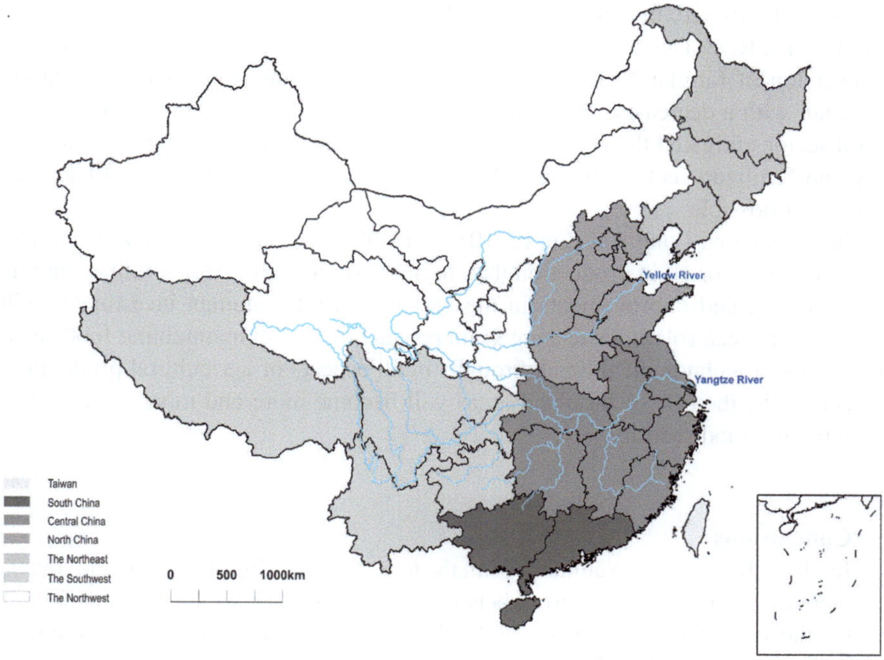


**Fig. 10.7** Division of China's provinces for Long and Zou's study (2010) (Source: Long and Zou 2010)

**Table 10.4** Indicator system for farmland use assessment

| Rule layer factors | Indicator layer factors | Definition |
|---|---|---|
| Investment intensity | Power | Gross power of farming mechanism per ha |
| | Labor | Gross farming labors per ha |
| | Fertilizer | Gross fertilizer utilization per ha |
| Management level | Multi-cropping index | Dividing the crop area by the area of farmland |
| | Irrigation index | Dividing the irrigated farmland area by the area of farmland |
| | Grain-farmland index | The proportion of grain-crop area in the total crop area |
| Production effect | Grain yield per ha | – |
| | Farming output value per ha | – |
| | Grain output per capita | – |

Source: Long and Zou (2010)

management level, and production effect, while taking farmland characteristics into consideration. In total, Long and Zou (2010) used nine indicators, three from each layer (Table 10.4).

Long and Zou (2010) compared the results of 1978, 1985, 1995, and 2004, and concluded that during the period 1995 to 2004 there were obvious improvements in the economy of each province. However, the impact of this economic development on the FUL differed. In regions with rapid development, such as Beijing and Tianjin, the Yangtze River Delta, and the southeast coastal provinces (i.e. regions with rapid conversion of farmland to non-agricultural activities, and transfer of farm labor, together with a decreasing grain output), the structural adjustments of the agricultural sector mitigated the positive influence that increasing investments in machinery and fertilizers had on the FUL. In these areas, improvements in the FUL were no longer possible.

In conclusion, Long and Zou (2010) argued that it is more important to protect farmland area for grain production than to increase the FUL. Along with economic development and improvements in the agricultural base, human investments will play only a weak role in increasing grain production and in maintaining food security, without technological breakthroughs in all aspects of agricultural production. Accordingly, the role of farmland areas will become more and more important in maintaining food security.

**Conclusions**

In this chapter we examined China's loss of agricultural output and the increase in prices that occurred between 1999 and 2003, which were blamed on the considerable amounts of land retired from agricultural production through the GfG. This led to a slow down in implementation of the program after 2004 (as discussed in Chap. 3). However, much of the literature reviewed in this chapter argues that, overall, the GfG played only a minor role in the decline in agricultural output. Overall, the slowdown was due to insufficient investment in the agricultural sector, a shift from the production of grain to the production of other food crops, and urban areas encroaching on highly productive agricultural land, particularly in the eastern provinces, where the GfG was not implemented. Indeed, while it is true that some counties experienced localized reductions in total production, in many other areas the GfG also led to increasing output because farmers could invest more money and time in their more productive fields. Nevertheless, concern for the future availability of agricultural products in particular isolated areas may be justified, especially where transportation of food can be problematic because of underdeveloped infrastructure.

# Chapter 11
# Participants' Income Levels

**Abstract** This chapter discusses the impact of the Grain for Green on the sources and level of income of farmers. While in many places agricultural incomes tended to dominate before the Grain for Green was introduced, by relieving farmers from agricultural work, the Grain for Green has had a considerable impacts on the economic structure and potential sources of income. With the Grain for Green the income structure diversified, to include agriculture, Grain for Green subsidies, the sale of Grain for Green-sponsored forest products, off-farm work in the villages of residence, and migration. In terms of the incomes from Grain for Green-induced land use changes, a distinction has to be made between economic trees, ecological trees, and grassland. Researchers agree that economic trees bring higher profits to the farmers, but even among economic trees, not all trees bring profits comparable to crops, once the subsidies are excluded from the calculation. Most researchers have looked at the benefits per hectare rather than the benefits per person-day.

**Keywords** Total incomes • Economic restructuring • Agricultural production • Sale of forest products • Off-farm work • Migration • Household composition • Income inequality

## Introduction

In China the majority of the rural poor are concentrated in resource deficient, remote, upland or mountainous, and sometimes minority-inhabited areas in the north, northwest and southwest. Although these poor have land use rights, in many cases the land itself is of such low quality that it is not possible to produce sufficient food for subsistence. Since the Grain for Green program targets low productivity lands in mountainous areas, it was implemented primarily among the rural poor, and consequently is expected to have a positive impact on China's efforts to reduce poverty. In this chapter we examine changes in the levels of income among the households that participated in the GfG. It is reasonable to expect that the generous subsidies, the labor freed from agricultural production, and the new opportunities given by the GfG and other programs in the targeted areas would promote the development of the local economies. The research, however, has produced conflicting conclusions.



C.O. Delang, Z. Yuan, *China's Grain for Green Program*,
DOI 10.1007/978-3-319-11505-4_11

The first section discusses changes in participants' total income and in income composition. Most researchers have found that participants' incomes have increased. However, the studies reviewed in this first section the participants to the non-participants. This is not sufficient, because increased incomes and asset values are not necessarily attributable solely to the GfG program; it is very likely that the incomes of non-participants also increased, and that the increased income of GfG participants is not entirely attributable to the GfG. The second section addresses this issue, by comparing incomes of participants to those of non-participants.

## Changes in Total Incomes

In 2000, Uchida et al. (2005) carried out a survey among 144 participating households from 16 randomly selected villages in Ningxia and Guizhou Provinces, and found that average household real net income increased after participating in the GfG program.[1] In Ningxia, from 1999 to 2000 the average real net household income increased 75 %, from Yuan 2,694 to 4,728. During the same period in Guizhou, it increased by 8 %, from Yuan 3,691 to 3,969. However, judging from the income structure change from 1995 to 2000 (Fig. 11.1), it seems that most of the increase from 1999 to 2000 was due to GfG payment. The differences in program payments, which themselves reflect differences in land holdings and participation in the program, explain most of the inter-provincial differences in income increases (Uchida et al. 2005).

Peng et al. (2007) looked at participants' net income in Zhangye City, a 41,924 $km^2$ prefectural-level administrative area at the center of the Hexi Corridor in western Gansu Province.[2] In 2001, it had a population of 1.26 million, of which about 1 million people (81.7 %) were involved in agriculture. Implementation of the GfG project in Zhangye began in 2002, and during the next 2 years, 286 $km^2$ of agricultural lands were converted into forestlands. Peng et al. (2007) assessed the costs and benefits to peasants engaged in the project to determine whether peasants benefited from participating in the project. Peng et al. (2007) found that the GfG had a positive impact overall on participants' net income in Zhangye city after 3 years of implementation. Except for new GfG participants in 2004, the net income of participating households was positive and increased over time (Table 11.1). The loss in 2004 might have been caused by a sudden policy change from expansion to forest/grassland maintenance (as discussed in Chap. 3).

The composition of total income shows the importance of the GfG in this particular region: between 2002 and 2004 household income was made up primarily of government subsidies (49.15 %) and migrant workers' income (40.10 %). Other

[1] The researchers collected information on households' on-farm production activities on a plot by plot basis. For each plot, respondents reported the crop(s) grown, yield, total output and inputs in 1999 before the program started. The survey also asked for detailed information on each household's total asset holdings and other income-earning activities from both on- and off-farm enterprises after the program began (Uchida et al. 2005).

[2] The survey was carried out in 2004 and included 313 randomly selected households from 13 villages.

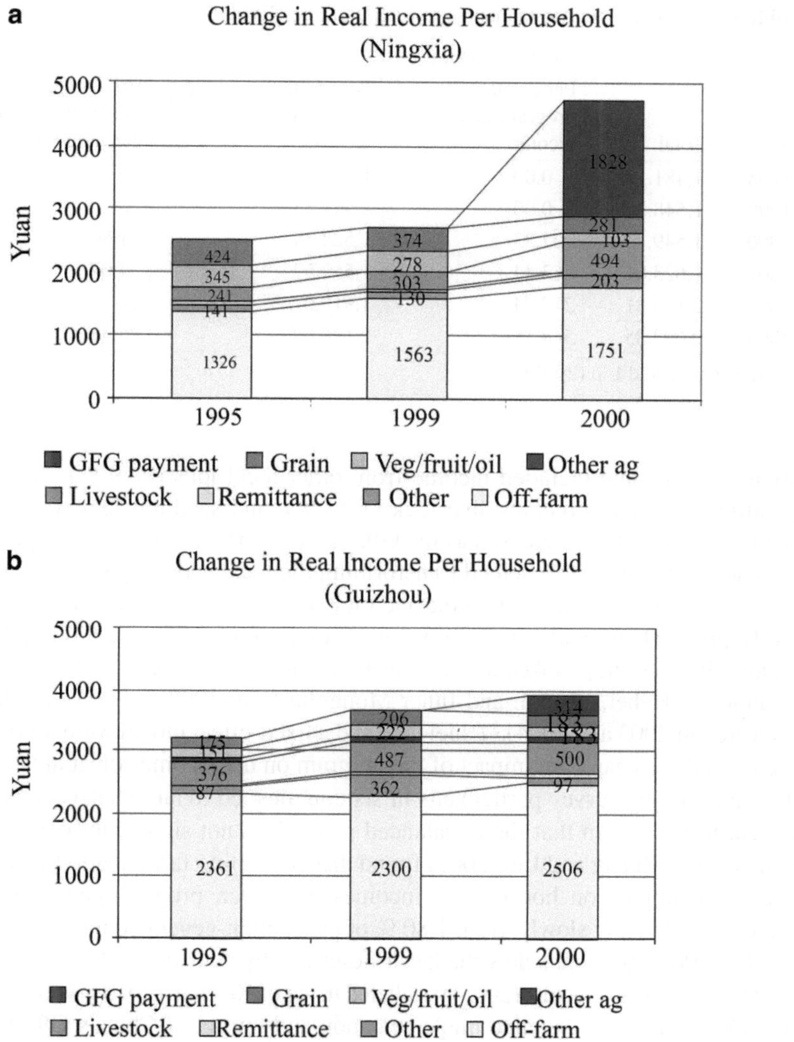


**Fig. 11.1** Changes in real income per capita of farm households participating in GFG in Ningxia and Guizhou provinces, 1995–2000 (Note: Data have been adjusted for inflation. Source: Uchida et al. 2005)

**Table 11.1** Net household income derived from the implementation of the GfG project in Zhangye City (2002–2004)

| | Net income (million Yuan) | | |
|---|---|---|---|
| Year of participation | 2002 | 2003 | 2004 |
| 2002 | 34.03 | 47.36 | 51.53 |
| 2003 | | 11.68 | 55.40 |
| 2004 | | | –9.21 |

Source: Peng et al. (2007)

**Table 11.2** Average per capita income in 17 counties in Hebei, Shanxi, and Inner Mongolia

| Year | Per capita net income (Yuan) | | | |
|---|---|---|---|---|
| | Total | Per capita program-generated income | Per capita household production-generated income | Percentage of program- to production-generated income |
| 1998 | 1,481.32 | 0.00 | 1,481.32 | 0 |
| 1999 | 1,548.55 | 0.00 | 1,548.55 | 0 |
| 2000 | 1,549.73 | 27.33 | 1,522.39 | 1.80 |
| 2001 | 1,623.01 | 63.44 | 1,559.57 | 4.07 |
| 2002 | 2,136.81 | 217.74 | 1,919.07 | 11.35 |
| 2003 | 2,692.85 | 544.18 | 2,148.67 | 25.33 |

Source: Zhang and Liu (2005)

sources of income included income from other local jobs (9.29 %), income from planting grass and breeding livestock (1.27 %), and seedling fees (0.19 %). It is likely that migration increased as the GfG freed labor from agricultural production, so it seems that the GfG helped transforming the local economy (Peng et al. 2007).

This case may be rather extreme and may be so because the region was comparatively poor. Zhang and Liu (2005), for example, found that the GfG had a much smaller impact; they looked at the contribution the GfG made to total incomes in 17 counties in Hebei, Shanxi, and Inner Mongolia from 1998 to 2003 (the data were collected in 2003 and 2004).[3] Panel data and a fixed-effect model were used to assess the immediate/near-term impact of the program on the incomes of rural households. Though missing survey participants in six counties led to an unbalanced panel, statistical tests confirm that the unbalanced panel does not significantly alter the ultimate results. Zhang and Liu (2005) found that converting farmland to forestland had a positive impact on households' incomes. However, program-generated income increased relatively slowly, from 1.80 % of production-generated income in 2000, to 25 % in 2003, still well below the levels described by Peng et al. (2007) (Table 11.2).

Zhang and Liu's results (2005) paralleled those of Xu et al. (2010), who used a 2003 household survey to examine implementation and impact of China's GfG Program.[4] Using a treatment effects approach to evaluate program impact, they found evidence of

[3] Liu and Zhang (2006) use data obtained from a unique panel survey conducted by the MOF in 17 counties of North China from 1998 to 2003, supplemented with village and county-level survey data. The 17 counties were randomly selected from 68 program-targeted counties in Hebei, Shanxi, and Inner Mongolia. A total of 188 households were sampled from the selected villages with a total of 927 observations.

[4] The data come from a household and village-level survey completed in 2003 by the Center for Chinese Agricultural Policy (CCAP), Chinese Academy of Sciences. The survey was conducted in the three provinces in which the GfG was first implemented, located at the upper reaches of the Yellow River Basin and the Yangtze River Basin: Shaanxi, Gansu, and Sichuan. Two counties per province, three townships per county, two participating villages per township, and 10 households per village were randomly selected, for a total of 36 village surveys and 360 household surveys (Xu et al. 2010).

positive impact on cropping, husbandry, and total income, though the results were not robust enough to support government claims of huge gains (Xu et al. 2010).

The high reliance of GfG subsidies found by Uchida et al. (2005) was not confirmed by Shi and Wang (2011). Shi and Wang (2011) conducted a long-period economic assessment that aimed to identify changes in the income structure of rural households 10 years after the project began. The fieldwork for that research was done in Mizhi County (in the northern part of Shaanxi Province).[5] The county covers 1,212 $km^2$ with an altitude ranging from 843 to 1,252 m, and is semi-arid with a middle temperate, continental climate. From an economic point of view Mizhi County is predominantly agropastoral, with more than 80 % of its total area being cropland, and its farmers raising a great number of goats. The county has 15 townships with 396 administrative villages and a population of more than 200,000, of which 180,000 are rural residents.

The GfG was implemented in all 15 townships of Mizhi County beginning in 1999. According to the Mizhi County Forestry Bureau, a total of 931.2 ha of croplands and degraded slope lands were converted to forestlands. This resulted in a dramatic increase in its farmers' per-capita net income. Statistical analysis showed that 9 % of the farm households increased their net incomes less than twofold, 69 % of the farm households increased their net incomes two to five fold, and 19 % increased their net incomes six to nine fold (Fig. 11.2 and Table 11.3). The per capita net income for farmers in Mizhi County increased by 317 % between 1998

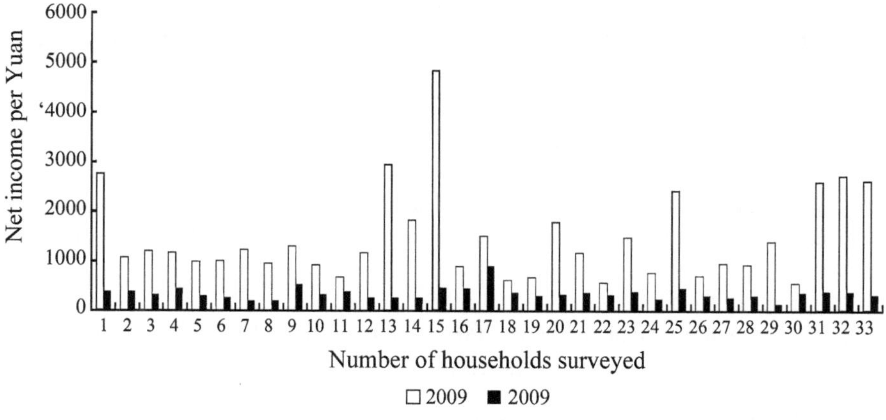


**Fig. 11.2** Net income of farm households before and after implementation of the GfG project (Source: Shi and Wang 2011)

[5]A questionnaire survey was adopted to investigate farm households in 2010. Villages were randomly chosen from each district in proportion to its area size. In total there were 33 valid samples. The study also employed other approaches to obtain data, including face-to-face interviews and informal discussions with local leaders/officials, group debate with local people and comments in official records about environmental policy. Based on Bossel (1999) social sustainability indicators, the social impact of the GfG project was assessed using the coordination coefficient in systems.

**Table 11.3** Overall net income and net income from crop planting of households before and after the GfG

| Before the GfG | | 10 years after the GfG | |
|---|---|---|---|
| **Net income (Yuan)** | **Percentage** | **Net income (Yuan)** | **Percentage** |
| 1,000–2,000 | 9.0 | 2,000–5,000 | 22.0 |
| 2,000–4,000 | 75.6 | 5,000–10,000 | 37.4 |
| 4,000–7,000 | 15.4 | 10,000–20,000 | 40.6 |
| **Net income from crop planting (Yuan)** | | **Net income from crop planting (Yuan)** | |
| 1,000–2,000 | 9.0 | 1,000–5,000 | 13.0 |
| 2,000–4,000 | 15.0 | 5,000–7,000 | 50.0 |
| 4,000–7,000 | 76.0 | 7,000–10,000 | 28.0 |
| | | >10,000 | 9.0 |

Source: Shi and Wang (2011)

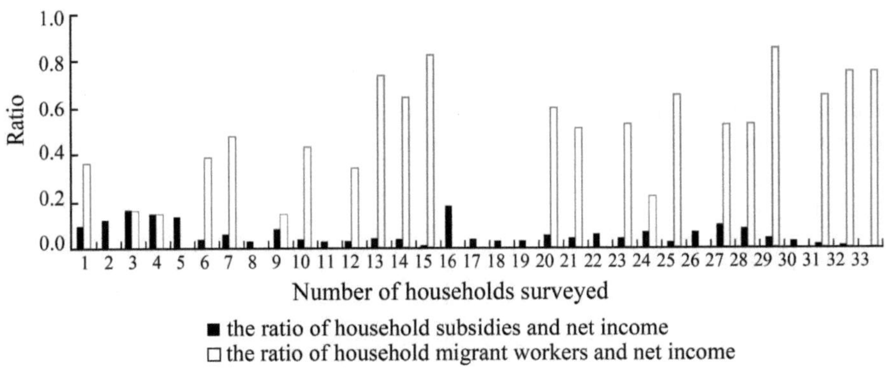


**Fig. 11.3** Proportions of government subsidies and incomes of the farm household members as migrant workers to the net incomes of the farm households (Source: Shi and Wang 2011)

and 2008, reaching Yuan 3,368 in 2008 (Anonymous 1998, 2002). The possible sources of the net income increases were:

1. Expanded animal husbandry and orchards;
2. Increased income of farmers from labor service outside their hometowns;
3. Increased prices of agro-products such as potatoes and other major farm produce.

Shi and Wang (2011) found that after losing cropland, a high proportion of the farm households had members who migrated to work in urban areas or became involved in other economic sectors locally. Hence, the incomes of migrant workers made the greatest contribution to household incomes (Fig. 11.3), far outpacing household subsidies. Almost 50 % of the farm households received more than 50 % of their net incomes from migrant workers.

**Table 11.4** Composition and structural change of household incomes over time (unit: Yuan in 1994 constant price)

| | Quantity | | | | | Percentage | |
|---|---|---|---|---|---|---|---|
| Year | Total | Agriculture | Off-farm | Subsidy | Other | Agriculture | Off-farm |
| **Shaanxi** | | | | | | | |
| 1999 | 3,848.0 | 2,413.7 | 1,108.3 | 326.0 | | 0.63 | 0.29 |
| 2000 | 4,375.6 | 2,533.4 | 1,320.9 | 521.3 | | 0.58 | 0.30 |
| 2001 | 4,501.7 | 2,566.9 | 1,426.7 | 508.1 | | 0.57 | 0.32 |
| 2002 | 5,187.8 | 2,653.5 | 1,714.6 | 819.8 | | 0.51 | 0.33 |
| 2003 | 5,400.5 | 2,458.7 | 1,739.8 | 1,201.9 | | 0.46 | 0.32 |
| 2004 | 6,091.3 | 2,688.0 | 1,863.7 | 1,539.5 | | 0.44 | 0.31 |
| 2005 | 7,290.6 | 2,388.5 | 2,764.4 | 1,854.1 | 283.6 | 0.33 | 0.38 |
| 2006 | 8,205.9 | 2,819.3 | 3,163.5 | 1,928.8 | 294.3 | 0.34 | 0.39 |
| 2007 | 9,294.7 | 3,130.5 | 4,178.6 | 1,493.8 | 491.9 | 0.34 | 0.45 |
| 2008 | 9,825.4 | 2,880.7 | 4,589.9 | 1,783.6 | 571.2 | 0.29 | 0.47 |
| **Sichuan** | | | | | | | |
| 1999 | 4,951.2 | 3,108.2 | 1,762.4 | 80.5 | | 0.63 | 0.36 |
| 2000 | 5,580.3 | 3,217.5 | 2,111.0 | 251.8 | | 0.58 | 0.38 |
| 2001 | 5,948.2 | 3,286.7 | 2,380.0 | 281.5 | | 0.55 | 0.40 |
| 2002 | 6,591.0 | 3,439.6 | 2,747.9 | 403.5 | | 0.52 | 0.42 |
| 2003 | 7,196.1 | 3,616.9 | 3,053.4 | 525.8 | | 0.50 | 0.42 |
| 2004 | 7,709.0 | 3,881.8 | 3,261.4 | 565.8 | | 0.50 | 0.42 |
| 2005 | 7,570.0 | 3,427.4 | 3,163.3 | 723.4 | 255.9 | 0.45 | 0.42 |
| 2006 | 8,540.3 | 3,847.2 | 3,651.6 | 767.0 | 274.4 | 0.45 | 0.43 |
| 2007 | 11,571.5 | 5,070.5 | 5,616.2 | 594.8 | 290.0 | 0.44 | 0.49 |
| 2008 | 12,445.6 | 5,316.8 | 6,157.8 | 554.4 | 416.6 | 0.43 | 0.49 |

Source: Yin and Liu (2011)
Note: "Other" means local welfare compensation and assistance to the poor and disabled

Yin and Liu (2011) compiled a unique longitudinal dataset from multiple rounds of surveys. The dataset covers ten consecutive years (1999–2008), containing a large but slightly fluctuating number of households, from 1,251 to 1,461, in six counties of two representative provinces in western China, Shaanxi and Sichuan. As shown in Table 11.4, while income from agriculture as a whole increased from 1999 to 2008, its pace of growth was much slower than that of off-farm and off-village income. In Shaanxi, total household income increased from Yuan 3,849 in 1999 to Yuan 9,825 in 2008 (Yin and Liu 2011). The greatest contribution to this increase was made from off-farm and/or off-village income, which rose from Yuan 1,108 to Yuan 4,590 during the same period. On the other hand, agricultural incomes only increased by Yuan 480. In Sichuan, the situation was similar. Total incomes increased from Yuan 4,951 in 1999 to Yuan 12,446 in 2008, with off-farm and off-village income jumping from Yuan 1,762 to Yuan 6,158, and agricultural income increasing more moderately, from Yuan 3,108 to Yuan 5,317. Overall, the share of agricultural income declined to only 29 % in Shaanxi, and from 63 to 43 % in Sichuan (Yin and Liu 2011).

Due to their larger percentage of participation and greater amount of land enrollment, households in Shaanxi benefitted tremendously from participating in the GfG. On average, a household there received an annual subsidy of up to Yuan 1,929 in 2006, accounting for almost 23.5 % of its total income in that year. In contrast, Sichuan had a modest increase in both enrolled households and enrolled cropland. Even though households received Yuan 70 per year more subsidy per mu, they did not benefit as much as their counterparts in Shaanxi. The highest level of subsidy was Yuan 767 in 2006, equivalent to almost 9 % of the total household income in that year (Yin and Liu 2011).

Some researchers, however, found that some areas had also experienced a decline, rather than an increase in household income due to the GfG. One such case was Dunhua County in the hinterland of the Changbai Mountains, northeast China, studied by Wang and Maclaren (2011).[6] They carried out a dichotomous logistic regression analysis to test for a relationship between perceived change in family income due to participation in the GfG, and a range of independent variables, including socioeconomic characteristics, land characteristics, and motivations for participation (Wang and Maclaren 2011).[7]

Wang and Maclaren (2011) found that, at the household level, 58 % of the families involved in afforestation felt that their income had declined after the GfG began. The impact of the program on the net income of participating households and sources of income are shown in Table 11.5. The average net income of households in three of the eight townships studied declined after the implementation of the GfG, but overall there was a growth of 13 %.[8] Farmers in Guandi Township experienced the largest absolute and percentage decline (74.2 %) in net income. There was no significant difference ($p>0.05$) in net income between the plots set aside and those not set aside for the program. The single exception is Xianru Township, where net income from non-participating plots was more than double that from participating

[6] Dunhua County covers an area of 11,957 km$^2$ and has a total population of 480,000. According to the land use map of Dunhua County for the year 2000, forest lands covered 76.6 % of the territory, and farmlands 15.6 %. Slopes less than 5° accounted for 87 % of the total cropland area. Dunhua County has been the pilot site for several nationwide forest protection projects, including the NFPP (SFA 2005e). In 2000, the county was selected as a demonstration site for the GfG, and all of its 16 townships participate in the GfG program. Since 2000, 230,000 ha of land have been converted to forests (Wang and Maclaren 2011).

[7] Wang and Maclaren (2011) selected townships randomly. In each township, two villages were selected and within the two villages 20 respondents were chosen at random. The primary data came from 156 questionnaires and obtained information about income and changes in economic structure of the family before (1999) and after (2003) participating in the program, especially about economic crops, livestock raising and off-farm work. Besides the household survey, interviews with government officials of the Dunhua Forestry Bureau and other agencies were conducted to understand the historical and geographical context of society and the economy in Dunhua, and gain an overview of the progress of the program. Social and economic data of afforestation in Dunhua County were derived from statistical yearbooks, development reports by Dunhua governments, publications on local agriculture, soil, forest and historical development (Wang and Maclaren 2011).

[8] Incidentally, families who experienced a decrease in income were more likely to claim that the land conversion had been forced on them by government action. Peasant families with higher incomes and more economic resources to cope with change were associated with more positive perceptions of land use conversion (Wang and Maclaren 2011).

**Table 11.5** Impact of the GFG on net household income and sources of income in Dunhua County

| County | | Net annual income per capita (Yuan) | Income change | Source of income (Percentage) | | | |
|---|---|---|---|---|---|---|---|
| | | | | Agriculture | Off-farm | Livestock | Economic crop |
| Dashan | BP | 2,804 | −31 % | 79.9 | 12.8 | 7.3 | 11.8 |
| | AP | 1,924 | | 69.5 | 17.1 | 13.4 | 24.5 |
| Emu | BP | 1,980 | 9 % | 79.8 | 8.3 | 11.9 | 17.3 |
| | AP | 2,156 | | 62.7 | 17.4 | 19.9 | 22.7 |
| Guandi | BP | 2,482 | −13 % | 73.5 | 17.7 | 8.8 | 24.4 |
| | AP | 2,152 | | 73 | 16.7 | 10.3 | 35.4 |
| Heishi | BP | 1,966 | −15 % | 73.2 | 12.2 | 14.6 | 2.1 |
| | AP | 1,664 | | 59.4 | 15.7 | 24.9 | 3.5 |
| Hongshi | BP | 1,023 | 57 % | 83.5 | 11.2 | 5.3 | 21.1 |
| | AP | 1,606 | | 82.2 | 12.0 | 5.8 | 25.6 |
| Huangnihe | BP | 2,945 | 33 % | 78.7 | 17.8 | 3.5 | 61.9 |
| | AP | 3,931 | | 74.3 | 22 | 3.7 | 70.5 |
| Shaheyan | BP | 2,820 | 25 % | 69.8 | 27.3 | 2.9 | 18.6 |
| | AP | 3,532 | | 64.5 | 32.1 | 3.4 | 28.4 |
| Xianru | BP | 1,879 | 42 % | 73.7 | 11.9 | 14.4 | 37.6 |
| | AP | 2,668 | | 62.7 | 15.5 | 21.8 | 43.3 |
| Average | BP | 2,237 | 13 % | 76.5 | 14.9 | 8.6 | 24.4 |
| | AP | 2,454 | | 68.5 | 18.6 | 12.9 | 31.7 |

Source: Wang and Maclaren (2011)
Note: BP=before GfG (1999), AP=after GfG (2003)

plots (Wang and Maclaren 2011). Nevertheless, Wang and Maclaren (2011) show some level of restructuring of the local economy due to the introduction of the GfG.

Overall, household income is still dominated by agriculture, even though its importance declined after the GfG was introduced. Perhaps not surprisingly, incomes from agriculture overall dropped (though in some cases very little), since some agricultural land was retired (Table 11.5). On average, agricultural income accounted for about 76.5 % of total household income before the GfG was introduced and 68.5 % after. On the other hand, surprisingly, off-farm incomes increased only marginally, from 14.9 to 18.6 %, perhaps indicating that most of the opportunities available locally had already been taken and there was little migration outside the area. Livestock became an important player in some counties, such as Emu, Heishi and Xianru, although on average its contribution to total sources of income increased from 8.6 % to only 12.9 %. On the other hand, "economic crops" (the authors define them as "including tobacco, flax and other crops") increased considerably in a few counties. Dashan led the way, with economic crops more than doubling their contribution to households' total incomes. On average, the contribution of economic crops to total income also increased quite considerably, compared to the increase in other sources of income.

**Table 11.6** Comparison of cropland and income per capita

| Study Area | Province | Cropland per capita (ha) | | Income per capita (Yuan) | |
|---|---|---|---|---|---|
| | | Before the program | 2000 | Before the program | 2000 |
| Dingxi City | Gansu | 0.336 | 0.227 | 2,022 | 1,487 |
| Pengyang County | Ningxia | 0.460 | 0.184 | 1,118 | 1,134 |
| Heqing County | Yunnan | 0.100 | 0.068 | 1,672 | 1,921 |
| Dafang County | Guangzhou | 0.149 | 0.040 | 1,484 | 1,197 |
| Tianquan County | Sichuan | 0.127 | 0.023 | 3,106 | 8,646 |

Source: Xu and Cao (2002)

The diversity of findings should not be surprising, given the social and environmental heterogeneity of China. Indeed, Xu and Cao (2002), compared five counties with different levels of cropland and income per capita, in five provinces, and found considerable variation in the consequences of the GfG program. In three cases out of five, household income increased following land retirement, in spite of less land being available (Table 11.6). In two cases, however, the income from farming dropped after the introduction of the GfG (although lower income in some counties may have been due to delayed delivery of grain and cash subsidies). Meanwhile, "income from non-farming activities increased across all the counties surveyed, suggesting that the potential for structural adjustment – reducing slope farming and exploring non-farming opportunities simultaneously – does exist, and these new activities should benefit local people and lead to sustained environmental improvement" (Yin et al. 2005: 27).

## Comparison of Program Participants and Non-participants

Uchida et al. (2007) argued that only examining households that participated in the program is not sufficient, because an increased average income and asset value of the participating households is not necessarily attributable solely to the GfG program; it is very likely that the incomes of non-participants also increased, and that the increased income of GfG participants is not entirely attributable to the GfG. Participants' characteristics may also have contributed to income increases. Since officials did not implement the GfG program on the basis of a randomized experiment, it cannot be assumed that the selection bias was zero. To test the actual contribution of the GfG, and obtain a more unbiased estimate of the impact of the GfG program on income, asset holdings, and labor allocation, Uchida et al. (2007) set out to hold constant variables that may affect total incomes, but are unrelated to the GfG. To do this, they employed three approaches: propensity score matching

**Table 11.7** Estimated effects of the GfG on changes in income, labor allocation and asset holdings using three approaches, 1999–2002

| | *PSM* | *DD* | *DDM* |
|---|---|---|---|
| Dependent variable | Y(2002) | Y(2002)–Y(1999) | Y(2002)–Y(1999) |
| Income per capita (Yuan) | −11.36 | 88.19 | −11.36 |
| Crop income per capita (Yuan) | −172.21*** | −114.34*** | −167.14*** |
| Other agricultural income per capita (Yuan) | 171.99** | 180.56* | 168.02** |
| Livestock inventories | 180.00*** | 180.00*** | 220.02*** |

Source: Uchida et al. (2007)
Notes: (1) The estimates are adjusted for inflation
(2) * significant at 10 %;** significant at 5 %; *** significant at 1 %

method (PSM), difference-in-differences (DD), and difference-in-differences matching method (DDM) (Uchida et al. 2007).[9]

The results of cross-sectional PSM analysis, which compares the matched participating and non-participating households with similar probability of participation, reveal that the GfG had some positive effects on participating households (Table 11.7). Although there is no statistically significant effect on the household's total income per capita, the PSM results suggest that the program had a significant positive impact on other agricultural incomes (from livestock activities), which increased by Yuan 172. In contrast, crop income dropped by the same amount. In addition, house value and livestock inventory values of the participating households increased by Yuan 486 and Yuan 180, respectively. The estimates for these variables were statistically significant (Uchida et al. 2007).

The results that show only a marginal (or negligible) impact on income are consistent with findings in Xu et al. (2003). Using DD analysis, Xu et al. (2003) found that there was a negative impact on cropping incomes and a positive impact on incomes from subsidies. In contrast to Uchida et al. (2007), however, Xu et al. (2003)

[9] Uchida et al. (2007) is based on surveys carried out in 2003, and commissioned by China's MOF as part of their effort to evaluate the nation's GfG program after the third year of implementation. By that time, this was the only existing dataset that included both participating and non-participating households. From the three provinces that had been participating in the GFG since 2000 (Sichuan, Shaanxi and Gansu provinces), two counties in each province and three townships in each County were randomly selected. In each township, two participating villages were selected, and within each village, ten households were randomly selected. There was at least one household participating in the program in every village. A total of 359 households were interviewed (Uchida et al. 2007). In two of the 36 villages, all of the households interviewed were participating households. In total, 75 % of the households interviewed participated in the GfG program. The household survey employed a sampling strategy designed to collect data on a random sample of households in the program area. Enumerators collected information on the household's production activities on a plot-by-plot basis, as well as detailed information on each household's total asset holdings, its demographic make-up, and other income earning activities from both on- and off-farm (Uchida et al. 2007).

used a model that was restricted and unadjusted for other variables and did not examine the impact of the conservation set-aside program on household assets or labor allocation. Using different versions of the propensity score matching method and survey data from 360 households for 1999 and 2003, Uchida et al. (2007) found that the GfG had only moderate success in achieving its poverty alleviation goals. They also did not find strong evidence to support the expected finding that participating households shifted their efforts into off-farm wage earning or self-employed activities, unlike what was found by others.

The findings from the DD analysis suggest that the program has had a significant impact on several income categories and several asset categories when comparing participating with non-participating households (Table 11.7). While crop income decreased (significantly) by Yuan 114, other non-crop agricultural income (from livestock enterprises) increased by Yuan 181, offsetting the decrease in crop income. Although the estimates are not statistically significant, the point estimates for fixed productive assets and livestock inventories were Yuan 683 and Yuan 161, respectively (Uchida et al. 2007). In findings largely consistent with the DD and PSM analyses, the DDM analysis results demonstrate that other agricultural incomes and the value of livestock inventories are higher for participating households (Table 11.7). Incomes from non-cropping agricultural activities increased by Yuan 168, while livestock inventories also increased, by Yuan 220. On the other hand, crop income declined by Yuan 167, as expected from a cropland set aside program, and confirmed by other studies (Uchida et al. 2007).

Using DD, PSM, and DDM approaches with different models, Uchida et al. (2007) found that there were positive, although somewhat nuanced, effects on participating households. The strongest finding was that participants increased their non-cropping incomes and asset bases to offset the fall in cropping incomes. Since Uchida et al. (2007) used cash accounting methods to measure assets, the higher direct income effects that might be associated with participation in the program could be offset by lower realized incomes from families who chose to increase their livestock holdings. In other words, if accrual accounting methods had been used, there would have been higher incomes. Moreover, 3 years is too short a time to assess the impact of a program on more fundamental structural transformations (Uchida et al. 2007).

Uchida et al. (2007) pointed out that a land retirement program, like the GfG program, had two effects on household labor: a substitution effect and an income effect. With a substitution effect a household retiring its cropland would shift the labor freed by the program into other productive activities, such as on-farm activities on the household's remaining cropland, off-farm wage jobs, or self-employment. But households may not shift all of the free time created by participating in the program into productive activities because of an income effect: a farmer could reallocate the time saved from the program into leisure. Hence, whether or not we could expect increases in off-farm labor is theoretically indeterminate (Uchida et al. 2007). Furthermore, a lack of increase in the incomes of participants may not necessarily indicate that the program failed, since the incomes of participants may have dropped if they had not participated in the program. Thus, assuming that participants' incomes must increase for the GfG to be considered successful is not necessarily correct.

As mentioned at the beginning of this section, we should expect that, after implementation of the program, both income levels and the income structure of participants and non-participants should change. This is because the opportunities available to households in the village or county change as additional money reaches the area or simply because, with the passage of time, prices of goods and opportunities change. Therefore, simply looking at socio-economic changes among participants, and assuming that all changes are due to the GfG, are likely to provide misleading information. More reliable results are obtained when the changes that occur during a given period of time among participants and non-participants living in close proximity are analyzed.

Accordingly, some studies compared changes among participants and non-participants. Xu, Bennett et al. (2004) compared the situation in Shaanxi, Gansu and Sichuan,[10] and found that, between 1999 and 2003, the growth rates in average net income varied greatly across regions. In Shaanxi, incomes of participants and non-participants exhibited a very similar growth rate; in Gansu, participant incomes showed a slower increase than that of non-participants; in Sichuan, participant incomes grew more rapidly than that of their non-participating counterparts. Overall, however, Xu et al. (2004) showed that the impact of the GfG on participants' income was statistically insignificant.

Table 11.8 presents the 1999 and 2002 components of total income for participant and non-participant households by province (Shaanxi, Gansu and Sichuan). Since such numbers could be the result of factors unrelated to the implementation of the GfG, they used a first-differences model explaining change in household per capita net income between 1999 and 2002, to more rigorously estimate program impact on income (Xu et al. 2010).

These numbers suggest that the GfG has indeed induced a restructuring of agricultural production, in which participants have shifted relatively more of their inputs from cropping into husbandry. In Shaanxi Province, growth rates for cropping income were 35 % for non-participants but only 12 % for participants (including subsidies received). In Gansu, cropping incomes dropped by 26 % and 32 % (including subsidies), respectively, while in Sichuan cropping income declined by 30 % for both groups (Xu et al. 2010).

Conversely, growth rates for husbandry were higher for participants than for non-participants. In Shaanxi, average household per capita husbandry income for participants increased by more than 1,055 %, compared to only 183 % for non-participants. In Gansu, participants' husbandry income grew by 1,783 %, compared with only 600 % for non-participants, and in Sichuan these numbers were 837 % and 500 %, respectively (Xu et al. 2010). However, changes in total income between participants and non-participants were less systematic across regions. Xu et al. (2010) estimated that in Shaanxi total income (including subsidies received) increased by 41 % and 42 % for participants and non-participants, respectively; for Gansu these numbers were 2.3 % and 12 %, respectively; and for Sichuan they were 26 % and 17 %, respectively (Xu et al. 2010).

[10] It is worth mentioning that, while the sample provinces in Uchida et al. (2005) and Xu et al. (2004) overlap, they studied different counties.

**Table 11.8** Per capita net income of participant and non-participant households, 1999 and 2002

| | Non-participant Households | | | | Participant Households | | | |
|---|---|---|---|---|---|---|---|---|
| | 1999 | | 2002 | | 1999 | | 2002 | |
| Income Component | Mean | Std | Mean | Std | Mean | Std | Mean | Std |
| **Shaanxi** | | | | | | | | |
| Total without subsidy | 940 | 777 | 1,335 | 930 | 986 | 1,077 | 1,325 | 1,874 |
| Total with subsidy | | | | | | | 1,394 | 1,877 |
| Cropping without subsidy | 465 | 521 | 626 | 429 | 420 | 672 | 401 | 622 |
| Cropping with subsidy | | | | | | | 470 | 628 |
| Husbandry | 6 | 23 | 17 | 63 | 18 | 78 | 208 | 916 |
| Off-farm | 388 | 623 | 590 | 947 | 401 | 554 | 525 | 680 |
| Other | 82 | 233 | 101 | 234 | 147 | 686 | 191 | 826 |
| **Gansu** | | | | | | | | |
| Total without subsidy | 1,803 | 1,681 | 2,021 | 1,741 | 1,287 | 980 | 1,287 | 942 |
| Total with subsidy | | | | | | | 1,317 | 942 |
| Cropping without subsidy | 484 | 350 | 360 | 246 | 589 | 523 | 370 | 320 |
| Cropping with subsidy | | | | | | | 399 | 345 |
| Husbandry | 17 | 53 | 119 | 220 | 6 | 30 | 113 | 222 |
| Off-farm | 1,192 | 1,570 | 1,346 | 1,624 | 633 | 679 | 681 | 647 |
| Other | 110 | 515 | 196 | 541 | 59 | 204 | 124 | 393 |
| **Sichuan** | | | | | | | | |
| Total without subsidy | 1,419 | 1,425 | 1,654 | 1,271 | | 1,195 | 1,961 | 1,524 |
| Total with subsidy | | | | | | | 2,067 | 1,151 |
| Cropping without subsidy | 721 | 938 | 506 | 633 | 829 | 931 | 472 | 590 |
| Cropping with subsidy | | | | | | | 577 | 583 |
| Husbandry | 33 | 42 | 202 | 200 | 49 | 75 | 459 | 1,187 |
| Off-farm | 543 | 953 | 714 | 987 | 674 | 897 | 869 | 971 |
| Other | 122 | 295 | 232 | 476 | 83 | 251 | 161 | 375 |

Source: Xu et al. (2010)
Note: All units are in 1999 Yuan, adjusted using the Rural Consumer Price Index

Yao et al. (2010) suggested that, in addition to the participation status and household characteristics, the impact of the GfG on income growth and labor transfer was determined by local economic development, program extent, and political leadership; further, the impact on income could vary from sector to sector. In other words, implementing the GfG could result in quite different outcomes for farming and animal husbandry, and thus total income, in different areas; it was likely that the program made a greater impact where a better developed economy, a larger program range, and stronger political leadership existed. This was one of the first studies that attempted to incorporate both internal and external variables, in studying the program's implementation and assessing its impact on rural economies (Yao et al. 2010).

To test these propositions, Yao et al. (2010) selected three counties in the Loess Plateau region: Wuqi (Yan'an municipality, Shaanxi province), Dingbian (Yulin municipality, Shaanxi province), and Huachi (Qingyang municipality, Gansu province), then looked at the changes that had occurred between 1999 and 2006. The rationale for this selection was as follows. First, these three counties represented the typical ecological conditions found in the region, where land degradation and soil erosion were so severe that there had been a great need for farmland retirement and conversion. Second, their adjacent locations and similar landscapes as well as program implementation schedules (they all initiated farmland conversion in the late 1990s, which was virtually completed by 2005) were conducive to comparisons. Third, while these counties are adjacent, they belong to different jurisdictions, which better reflect the variations in program execution, political setting, and economic development (Yao et al. 2010).

Situated in the northeast of Yan'an municipality, Wuqi has a total population of 127,369, with 109,470 rural residents. Unlike its neighbors, the county, which has been extremely poor, has enjoyed preferential treatment by the central government for two reasons. First, it has major oil and gas reserves. Second, it occupies a significant place in contemporary Chinese history as the end point of the Red Army's Long March (Wuqi GfG office 2007). This attention, since the mid-1980s, has enabled Wuqi's economy to grow rapidly. In 2005, the county's GDP was Yuan 2.1 billion, and its own revenue reached Yuan 0.7 billion. In recent years, Wuqi has become one of the richest counties in western China (Wuqi Statistics Bureau 2006).

Before 1998, Wuqi had 123,700 ha of cultivated land, or 3.40 ha per household, and a large number of the rural households also raised goats – the number of goats peaked at 280,000. As a consequence of extensive farming and open grazing, the county's land and vegetation were heavily degraded, causing severe water runoff and soil erosion. In response, Wuqi began retiring croplands on steep slopes and converting them to forest and grass coverage in 1998. Taking advantage of the GfG, Wuqi's land set-aside and conversion expanded tremendously in 1999. Croplands were cut back to 10,000 ha, and open grazing was banned in favor of raising goats in pens and vegetation recovery (Wuqi Statistics Bureau 2006, in Yao et al. 2010). By 2006, over 97,000 ha of converted cropland had passed the national survival, growth, and stocking inspections (Wuqi GfG Office 2007, in Yao et al. 2010).

Lying in the transitional zone between the Loess Plateau and the Erdos Desert, Dingbian is located in the west part of Yulin. Over 87 % of its population of 315,851 lives in rural areas (Dingbian Statistics Bureau 2006). On the other hand, Huachi is located in the eastern part of Gansu province, and 86 % of its 130,175 population is rural (Huachi Statistics Bureau 2006). As with Wuqi, both counties are endowed with rich petroleum and gas resources, as well as extensive farming and open grazing. Dingbian and Huachi, however, have not been allowed to develop their natural resources. Instead, the national company Changqing Petro Co., holds exclusive rights to exploration. While the GDP of Dingbian and Huachi in 2005 was close to Yuan 3 billion and Yuan 4.6 billion, respectively, higher than that of Wuqi, much of the profits from oil and gas extraction were retained by the oil company, and did not significantly benefit the local treasury.

Consequently, the total budget for Dingbian and Huanchi counties was less than Yuan 60 million in 2005 (Dingbian Statistics Bureau 2006; Huachi Statistics Bureau 2006, in Yao et al. 2010).

The total amount of cropland retired through the GfG by 2006 was 10,966 ha for Huachi and 21,905 ha for Dingbian, suggesting a much smaller extent of program implementation, given their total cropland holding in 1997 of 57,265 ha and 83,333 ha, respectively. Further, extensive farming and open grazing in these two counties were still the norm rather than the exception. Their local investment in land retirement was negligible, and incidences of delayed delivery and deduction of farmers' subsidies occurred (Dingbian GFG office 2007; Huachi GfG Office 2007 Compared to Wuqi, Dingbian and Huachi lacked political leadership, local investment, and extensive participation (Yao et al. 2010).

In August 2007, Yao et al. (2010) conducted a survey of 200 randomly chosen households in each of the three counties, including basic household characteristics, production, consumption, income, and farmland retirement and conversion. The data revealed that there was little difference in the number of laborers, the average amount of education in years, and the average age of household head between participating and non-participating households. On the other hand, noticeable differences existed in family size, cultivated land, and years of schooling of household heads (Yao et al. 2010).

Dividing income into discrete categories that included farming, animal husbandry, off-farm work,[11] and other sources, enabled Yao et al. (2010) to look into the gains and losses experienced by different sectors. They then used the DD model to detect the program's impact. Table 11.9 compares per capita income of the participant and the non-participant household groups in Wuqi in 1999 and 2006. Except for participating households' income from animal husbandry, all categories of income increased during the period under consideration. Both non-participants and participants saw their incomes from crop production increase, but non-participants saw their incomes increase by more: in 1999 the difference between non-participating households and participating households was Yuan 1,859, while in 2006 the difference had dropped to Yuan 1,136. Even though the amount of cultivated land of participating households was reduced, their improved productive efficiency seems to have reduced the income gap from crop production with non-participating households (Yao et al. 2010). Income from animal husbandry increased both for non-participating and participating households, but more so for participating ones. Similarly, income from off-farm employment increased for both participants and non-participants, but comparatively more for participating households. As mentioned, unlike other studies, Yao et al. (2010) found that even before the GfG was introduced in the region, participating households engaged more in off-farm incomes than non-participating households. Finally, other incomes increased by a similar amount for both groups. Overall, the income of participating households was much higher before implementation of the GfG, and remained equally higher in 2006 (Yao et al. 2010). These findings, again, differ from those of other studies.

[11] Off-farm employment includes (a) employment in local non-agricultural activities and (b) off-village employment as migratory workers.

**Table 11.9** Per capita average income of surveyed households in Wuqi, 1999 and 2006

| | Non-participating households | | Participating households | | Between group income difference | |
|---|---|---|---|---|---|---|
| Type of income[a] | 1999 | 2006 | 1999 | 2006 | 1999 | 2006 |
| Crop production income | 5,591 | 5,788 | 3,733 | 4,653 | 1,859** | 1,136 |
| Animal husbandry income | 1,162 | 1,948 | 3,575 | 1,409 | −2,413** | 539 |
| Off-farm income | 2,475 | 2,916 | 10,404 | 13,785 | −7,930 | −10,869*** |
| Other income | 0 | 5,411 | 61 | 6,778 | −61 | −1,367 |
| Total income | 9,228 | 16,064 | 17,773 | 26,625 | −8,544*** | −10,561*** |

Source: Yao et al. (2010)

Notes: (1) These statistics are rounded mean values, so they may not add up to the total exactly
(2) ***, ** represent significance levels of 1 % and 5 %, respectively

[a]Crop production income comes from producing corn, potatoes, and other minor crops; animal husbandry income comes from raising livestock, predominantly goats; off-farm income is comes from off-farm employment, mainly construction and service work in local towns as well as large cities; other income is derived from sources such as family properties and government subsidies; total income is the gross income from all sources (Yao et al. 2010)

**Table 11.10** Per capita average income of surveyed households in Huachi and Dingbian, 1999 and 2006

| | Non-participating households | | Participating households | | Between group income difference | |
|---|---|---|---|---|---|---|
| | 1999 | 2006 | 1999 | 2006 | 1999 | 2006 |
| Crop production income | 2,176 | 4,511 | 2,475 | 4,615 | −299 | −104 |
| Animal husbandry income | 2,371 | 1,591 | 1,358 | 1,265 | 1,012 | 326 |
| Off-farm income | 6,409 | 5,568 | 6,642 | 9,912 | −234 | −4,344 |
| Other income | 1,459 | 1,708 | 487 | 535 | 972*** | 1,172* |
| Total income | 12,414 | 13,379 | 11,962 | 16,327 | 1,452 | −2,948 |

Source: Yao et al. (2010)

Notes: (1) These statistics are rounded mean values, so they may not add up to the total exactly
(2) ***, * represent significance levels of 1 % and 10 %, respectively

Similar results have been found among the households surveyed in Huachi and Dingbian (Table 11.10); however, crop production incomes were slightly higher for participating households than for non-participating households in 1999, while in 2006 they were about the same. Since participating households had a considerable amount of farmland set aside, as in the case of Wuqi, we can conclude that they were able to considerably increase the productivity of their remaining land.

Differences can also be observed in incomes from animal husbandry. While in Wuqi the income from this activity for participating households had more than halved, in Huachi and Dingbian it dropped only marginally. Remarkably, in these two counties income from animal husbandry also dropped for non-participating households, unlike in Wuqi. Off-farm incomes dropped for non-participating farmers, perhaps

**Table 11.11** Regression results of income and off-farm employment based on the model with specific variables for regional variation

| | Crop production income | Animal husbandry income | Off-farm income | Other income | Off-farm employment | Total income |
|---|---|---|---|---|---|---|
| Status of participation | 131.11 | −2,445.52 | 3,170.06 | 382.16 | 0.09 | 5,397.04 |
| Economic condition | 619.27 | 202.64 | 187.94 | −269.32 | 0.25 | 286.52 |
| Program extent | 170.25 | 73.69 | 62.95 | −145.46 | 0.12 | 175.97 |
| Political leadership | 251.33 | 68.18 | 55.18 | −50.79 | 0.07 | 91.63 |
| Education of household head | 83.55 | 191.92 | 522.17 | 138.29 | 0.02 | 1,059.97 |
| Family size | 8.37 | 507.66 | 191.12 | 1,309.85 | 0.14 | 1,867.99 |
| Number of laborers | 190.59 | 258.93 | −1,792.95 | −498.13 | 0.07 | 1,376.97 |
| Non-agricultural employment | 187.41 | −606.91 | 9,191.11 | 126.79 | *NA* | 11,046.1 |
| Per capita cultivated land | 984.56 | −159.15 | −328.14 | 252.31 | −0.02 | 231.62 |
| $R^2$ | 0.58 | 0.40 | 0.25 | 0.20 | 0.48 | 0.15 |

Source: Yao et al. (2010)
*NA:* The non-agricultural employment variable is not included in the off-farm employment transfer model

because they concentrated more of their labor on crop production. Like the situation in Wuqi, however, off-farm incomes increased for farmers in Huachi and Dingbian. The result was that incomes increased only marginally for non-participating households, but they increased considerably for participating ones. Unlike in Wuqi, in 1999 participating households had slightly lower incomes than non-participating households. By 2006, however, as in Wuqi, non-participating households had been able to reverse the situation and had higher incomes than participating ones.

Table 11.11 shows the econometric relationship between various sources of income (first column) after the implementation of the GfG, and different variables (first row). The data reveal the following: First, all of the variables have a positive effect on the crop production income regression. Compared to non-participating households, crop production income of participating households increased by Yuan 131.1, which is not a large amount but is significant at the 99 % level. A better-developed local economy, a larger program, and stronger political leadership, respectively, result in an increase in the households' crop production income by Yuan 619.3, 170.2, and 251.3 at the 99 % significance level. Together, these add up to a significant increase (Yuan 1,240), partially confirming what Yao et al. (2010) hypothesized: variations in local programmatic, economic, and political conditions all affect crop production income. The head of household's education level also has a significant influence on crop production income, with each additional year of schooling leading to an increase of Yuan 83.6. Other variables, like the number of household laborers, per capita cultivated area, and non-agricultural employment, also led to a significant increase in crop production income (Yao et al. 2010).

Second, the regression of animal husbandry income reveals that participation status is negatively associated with the income at the 95 % significance level. Animal husbandry income of participating households is decreased by Yuan 2,445.5, in comparison to that of non-participating households (Yao et al. 2010). Here, program extent, economic development, and political leadership do not matter much. Variables like years of schooling for household head, family size, and number of household laborers have a positive but statistically insignificant effect. Likewise, per capita cultivated area and local non-agricultural employment have a negative but statistically insignificant effect (Yao et al. 2010).

Third, the off-farm income is positively related to the participation status and household head's years of schooling at the 90 % significance level. Participation allowed a household's off-farm income to increase by Yuan 3,170.1, and one additional year of schooling for the household head led to an increase of Yuan 522.2. Local economic development, program extent, and political leadership caused household off-farm income to increase respectively, by Yuan 187.9, 62.9, and 55.2. These effects are all significant at the 99 % level (Yao et al. 2010). Additionally, non-agricultural employment had a positive effect at the 99 % significance level. One more family member employed in the nonagricultural sector resulted in the household's off-farm income increasing by Yuan 9,191.1. In contrast, family size, number of household laborers, and per capita cultivated area did not have strong correlations with off-farm income. As to income from other sources, the regression has only one significant variable, family size, suggesting that the larger the family, the greater the income. All the other variables have little effect (Yao et al. 2010).

Fourth, the regression of the number of off-farm employment revealed that participation had a positive effect on off-farm employment at the 95 % significance level. Other things being equal, participation caused 0.09 units of labor to shift out (Yao et al. 2010). Although there was a positive relation with years of schooling for household heads, this relation was statistically insignificant. While family size and number of household laborers had positive effects on off-farm employment, per capita cultivated area had a negative effect on off-farm employment. These results illustrate that the more surplus labor a family has, the more off-farm income it generates; and, the larger the per-person cultivated area, the less likely it is for the household to engage in intensive farming, making it more difficult to seek off-farm work (Yao et al. 2010). Local economic development has a positive relation with off-farm employment; a coefficient of 0.25 indicates that the condition is a key factor in labor transfer. Program extent has an effect of 0.12, and political leadership has an effect of 0.07. Together, these variables cause 0.45 units of labor to shift out of farming, which is more than four times greater than the coefficient of participation status alone. This has further confirmed the hypothesis of Yao et al. (2010): the realized transfer of surplus farming labor depends on both internal and external conditions, coupled with program participation (Yao et al. 2010).

Fifth, total income has a positive correlation with years of schooling for the household head, family size, number of laborers, and non-agricultural employment. The contribution of these variables is Yuan 1,056 from one more year of household head education, Yuan 1,870 from one more person in the household, Yuan 1,377 from one more family laborer, and more substantially, Yuan 11,046 from one more non-agricultural job (Yao et al. 2010). Participation in the land conversion program results in an increase in total income by Yuan 5,397. In addition, local economic development, program extent, and political leadership are positively correlated with total income. Their coefficients, respectively, are Yuan 287, Yuan 176, and Yuan 91.6 (Yao et al. 2010).

Yao et al. (2010) conclude that, while the GfG has had a significant positive impact on crop production income, the magnitude of this effect is small. In comparison, better local economic conditions, larger program extent, and stronger political leadership have had a much greater impact. These results suggest that cropland retirement does not necessarily cause a reduction in crop yield or income if the production mode can be sufficiently transformed by adopting better inputs and management practices. With regard to income from animal husbandry, however, participation has had a substantial negative effect that is almost ten times greater than the combined positive effects of local economic conditions, program extent, and political leadership. Clearly, animal husbandry was hit hard by the grazing and feeding constraints imposed by the GfG, even with local efforts to maintain its vitality (Yao et al. 2010).

Meanwhile, participation has had a large positive effect on both off-farm income and total income. In combination the results indicate that, although animal husbandry has been negatively affected, the program's impact on other sectors has been positive therefore more than offsetting the aggregate negative effects (Yao et al. 2010). The results of the off-farm employment and income regressions highlight the fact that participating in the program has accelerated the transfer of farming labor and has greatly stimulated income growth from off-farm opportunities. Moreover, these positive effects have been reinforced by better economic development, larger program extent, and stronger political leadership. These findings are new and they have provided further supporting evidence for Yao et al.'s (2010) claim that the socioeconomic effects of the program are indeed predicated on the program's local range and conditions, coupled with participation status.

Yao et al. (2010) argue that while their findings regarding the program's negative effect on animal husbandry income for participants, and its positive effect on off-farm employment and total income, conform to what was previously reported (for example by Guo et al. 2005), confirming a positive effect on cropping income is new. This implies that cropland reduction will not inevitably cause a decline in crop yield, and thus income. Yao et al. (2010) propose that the significance of these effects is directly related to their sample features, including the selection of a representative study site, the coverage over a long period of time, the division of total income into specific categories, and the capture of specific regional variations. They conclude that the differences between their results and those of Xu et al. (2004) and Yi et al. (2006) lie in these factors.

Another study that compared participants and non-participants was carried out by Yin and Liu (2011) using a dataset that contained a large but slightly fluctuating number of households (from 1,251 to 1,461) in six counties of two representative provinces in western China – Shaanxi and Sichuan – over ten consecutive years (1999–2008). Table 11.12 disaggregates the net revenues from grain and livestock

**Table 11.12** Total and off-farm incomes for the two different groups of households in Sichuan (unit: Yuan in 1994 constant price)

| | Total Income | | Grain Production | | Livestock Production | | Off-farm Income | |
|---|---|---|---|---|---|---|---|---|
| Year | Participants | Non-participants | Participants | Non-participants | Participants | Non-participants | Participants | Non-participants |
| 1999 | 5,124.9 | 4,940.9 | 1,141.2 | 1,201.8 | 1,197.6 | 1,594.4 | 2,107.9 | 1,742.1 |
| 2000 | 6,334.0 | 5,302.1 | 1,147.8 | 1,222.7 | 1,705.8 | 1,639.5 | 2,397.4 | 2,005.3 |
| 2001 | 6,541.8 | 5,666.2 | 1,128.7 | 1,230.9 | 1,653.6 | 1,723.6 | 2,647.8 | 2,252.8 |
| 2002 | 7,050.6 | 6,202.2 | 1,144.5 | 1,294.5 | 1,680.5 | 1,880.0 | 3,078.7 | 2,468.2 |
| 2003 | 7,525.1 | 6,682.6 | 1,239.1 | 1,323.1 | 1,801.2 | 1,983.4 | 3,266.3 | 2,721.0 |
| 2004 | 8,136.2 | 6,991.6 | 1,266.9 | 1,402.8 | 1,988.6 | 2,143.5 | 3,580.1 | 2,726.1 |
| 2005 | 7,890.9 | 6,990.5 | 1,051.1 | 1,114.4 | 1,603.7 | 1,890.3 | 3,720.8 | 2,874.5 |
| 2006 | 8,907.6 | 7,866.2 | 1,117.3 | 1,191.3 | 1,839.9 | 2,243.1 | 4,294.6 | 3,249.5 |
| 2007 | 11,654.9 | 11,418.8 | 996.5 | 1,264.0 | 3,168.1 | 3,983.1 | 6,124.9 | 5,505.4 |
| 2008 | 12,127.3 | 12,965.8 | 907.2 | 1,180.6 | 3,022.7 | 3,819.6 | 6,555.1 | 6,606.0 |

Source: Yin and Liu (2011)

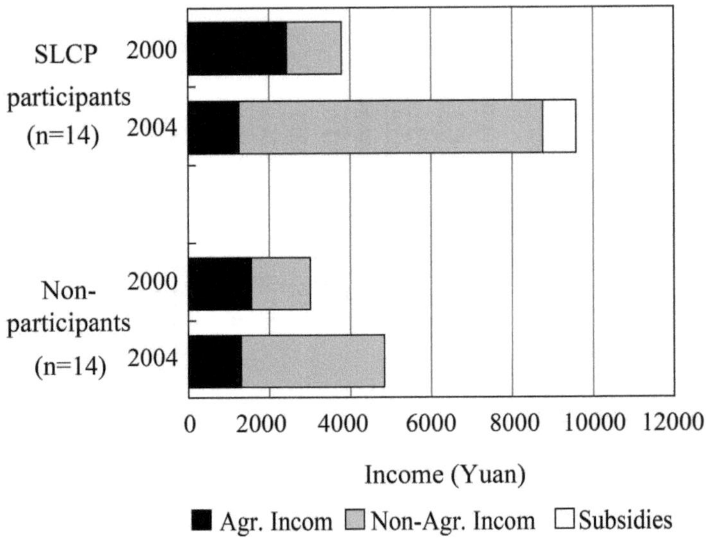


**Fig. 11.4** Income of participating and non- participating households in GfG, 2000 and 2004 (Source: Hori and Kojima 2008)

production in Sichuan for participating and non-participating households. Compared to participants, non-participants were able to maintain a slightly higher level of net revenue from grain and livestock production, although during the last 2 years of the survey (2007–2008) this difference increased. Still, off-farm incomes were much higher among participants, as expected, since participants were in some measure liberated from farm work. Overall, the higher on-farm income of participants and off-farm income of non-participants were approximately equivalent, and the difference in total income fluctuated around Yuan 1,000 in favor of participants.

Hori and Kojima (2008) addressed the differences between participants and non-participants in Mizhi County, Yulin City, Shaanxi Province, using a dataset of 27 participant and non-participant households from 2000 to 2004. They found that there were few differences in total incomes of participants and non-participants in 2000, with GfG participants earning slightly more than non-participants. These differences grew considerably by 2004, however (Fig. 11.4). Among participants, agricultural incomes declined about 50 % between 2000 and 2004, but non-agricultural incomes increased more than five-fold, jumping from Yuan 1,448 in 2000 to Yuan 7,559 in 2004. Non-agricultural incomes consisted mainly of incomes from migrant workers. The number of migrant workers among GfG participants increased considerably (Table 11.13). Overall, including GfG subsidies, incomes more than doubled from 2000 to 2004, with the income by migrant workers making up more than 70 % of nonagricultural income for both types of households in 2004. More than 50 % of villagers were working in Yulin City district. This can be attributed to the economic growth that occurred in Yulin thanks to the "Go West" campaign, which began in 1999 and can be said to have had a positive impact on

**Table 11.13** Variety of migrant workers

| Year | GfG participants | Non-participants | Manual labor | Others |
|---|---|---|---|---|
| Before 1998 | 2 | 4 | 5 | 1 |
| 1998 | 1 | 1 | 0 | 2 |
| 1999 | 2 | 0 | 1 | 1 |
| 2000 | 0 | 1 | 1 | 0 |
| 2001 | 0 | 2 | 2 | 0 |
| 2002 | 5 | 1 | 3 | 3 |
| 2003 | 12 | 1 | 5 | 8 |
| 2004 | 2 | 0 | 1 | 1 |
| **Total** | **24** | **10** | **18** | **16** |

Source: Hori and Kojima (2008)
Note: Data represent the result of interviews with householders, Unit: person

local migrant workers, thereby contributing to the success of the GfG. Nevertheless, the authors point out that Yulin is rich in mineral resources while most Chinese rural areas have fewer local opportunities, thus lower non-agricultural incomes. Meanwhile, among non-participants, while agricultural incomes decreased very slightly, nonagricultural incomes did not increase as much as those of GfG participants (Fig. 11.4). The result is that, while in 2000 GfG participants earned only slightly more than non-participants, by 2004 they earned twice as much.

## *Roles of Household Members and Importance of Household Composition*

Household composition, in particular the ages of household members, is important in determining the opportunities available to them beyond the village, and therefore the ways in which households adapt to reforestation programs. Liang et al. (2012) looked at the ways in which household composition determined the impact of the GfG program on household activities and livelihood in Zhouzhi County, one of the poorest counties in Shaanxi province, with Yuan 3,023 average per capita income in 2005. The county has a total area of 2,949 km$^2$, most of which is located in the Qinling Mountains, a natural boundary between northern and southern China. In 2002, Zhouzhi County introduced the GfG in the mountain towns.

Liang et al. (2012) is based on interviews and questionnaire surveys that used multiple level cluster sampling, conducted in April 2008.[12] Liang et al. (2012) first

[12] At the household level, cluster sampling was used for the questionnaire survey in 20 villages from the four selected towns. 1,078 questionnaires were completed, covering both participating and non-participating households with a variety of detailed information on demographic characteristics, production and consumption activities, income and other livelihood, as well as some basic information on each family member. In particular, the questionnaire addressed households' assets that did not change much even after participation in the program (Liang et al. 2012).

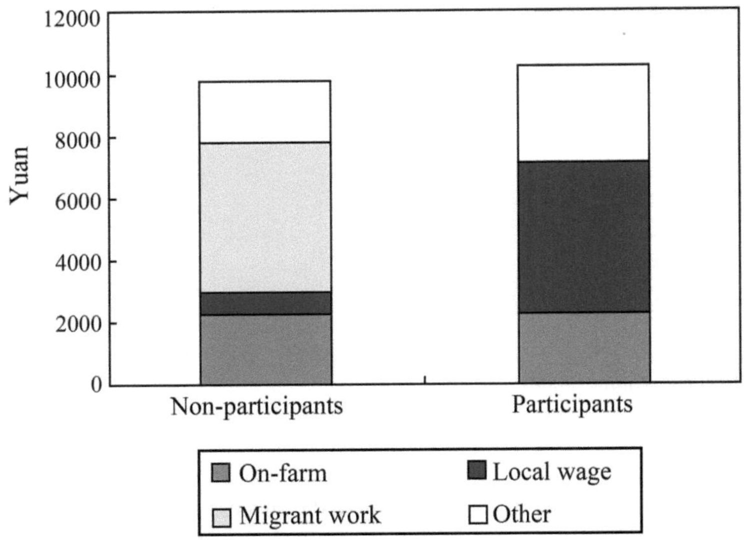


**Fig. 11.5** Mean household income portfolios in 2007 of households with children but no elderly (Source: Liang et al. 2012)

divided household members into three groups by age: children (<15 years old), adults (15–65), and elderly (>65). Due to mandatory education laws and a traditional emphasis on education, children are primarily in school. In order to test different policy effects dependent on various types of household composition, Liang et al. (2012) adopted the calculation formula of Uchida et al. (2007) to estimate household income.

The study found that: (1) participants who had children but no elderly had relatively more local wage income and less migratory wage income, due to lower propensity to migrate; (2) on-farm income was almost the same for the two groups (Fig. 11.5), because participants had less crop income but more forestry income after some of their land was converted; (3) participants had slightly more income than non-participants, since more often than not payments were more than the opportunity costs of the retired land; (4) notwithstanding overall significant positive effects on household local wage income, participating in the program had negative effects on migrating income; (5) local wage incomes were larger for participants than for non-participants. Non-participants earn a much larger share of household income from migration than from local wage work, while participants earned comparatively more from local wage work (Fig. 11.5); (6) estimation of income without payments showed that households with children but no elderly (H(C, A)) relied relatively more on payments from the GfG (Table 11.14). This can also shed light on which types of households will be most affected by cessations of payments from the program. The lower level of migration among GfG-participants may be caused by fewer opportunities for migration because of lower education levels and/or a weaker social network outside the locality. In other words, the GfG

**Table 11.14** Estimated impact of the Grain for Green Program and household composition on income

| | Total income[a] | On-farm income[b] | Income without payments[c] | Migrating wage-income[d] | Local wage-income[e] |
|---|---|---|---|---|---|
| Households with adults and elderly H(A,E) | −270.51 | −123.5 | −1,636.11 | 867.67 | 2,594.41 |
| Households with only adults H(A) | −743.39 | −347.76* | −1,862.29** | −276.77 | 2,354.03 |
| Households with children and adults H(C,A) | −1,895.91** | −568.75** | −3,329.07*** | −6,946.50*** | 5,859.25*** |
| Households with children, adults and elderly H(C,A,E) | 278.57 | 280.76 | −1,304.99 | −2,403.46 | 2,985.06 |

Source: Liang et al. (2012)

Notes: (1) [a,b,c]: Coefficients are estimated by OLS. [d,e]: Coefficients are estimated by Tobit

(2) *, **, *** indicate significance at 10 %, 5 % and 1 % levels, respectively

may not be able to promote migration as much as may be desirable, in spite of the financial incentives that people would have, because of social or structural constraints. It is quite possible that, without GfG subsidies, people would migrate even less, since there would be financial deterrents, such as loss of income from crop production (Liang et al. 2012).

Unlike areas where households can increase income from livestock activities and other types of assets after they participate in the GfG (as discussed, for example, by Uchida et al. 2007), Liang et al. (2012) found that villages' ecological policies were intensively implemented and households' activities were heavily restricted. For the H(C, A) group, which was heavily reliant on payments, if payments were removed from total income, participants could face a more serious income loss than other groups. Previous studies (Uchida et al. 2005; Xu et al. 2004) also found that, on average, increased income for participating farmers in Guizhou and Ningxia was due mainly to program payments (Liang et al. 2012). Other explanations for income loss from the program for the H(C, A) group are that the households were risk averse and took the subsidies as a risk-coping strategy, or a household over-anticipated the possibility of engaging in alternative, income-generating activities.

## *Income Inequality*

Because the GfG primarily targeted the least-productive land and the poorest households, it should be expected that the program contributed to decreasing inequality in the communities where it was implemented. Yin and Liu (2011)

**Table 11.15** Estimated Gini coefficients and their sources

| Year | Yuan in 1994 constant prices | | | | | |
|---|---|---|---|---|---|---|
| | Gini Coefficient | Total Income | Agriculture | Off-farm Income | Subsidies | Other |
| **Shaanxi** | | | | | | |
| 1999 | 0.34 | 3,848.0 | 2,413.7 | 1,108.3 | 326.0 | 0.0 |
| 2000 | 0.34 | 4,375.6 | 2,533.4 | 1,320.9 | 521.3 | 0.0 |
| 2001 | 0.34 | 4,501.7 | 2,566.9 | 1,426.7 | 508.1 | 0.0 |
| 2002 | 0.34 | 5,187.8 | 2,653.5 | 1,714.6 | 819.8 | 0.0 |
| 2003 | 0.34 | 5,400.5 | 2,458.7 | 1,739.8 | 1,201.9 | 0.0 |
| 2004 | 0.34 | 6,091.3 | 2,688.0 | 1,863.7 | 1,539.5 | 0.0 |
| 2005 | 0.28 | 7,290.6 | 2,388.5 | 2,764.4 | 1,854.1 | 283.6 |
| 2006 | 0.29 | 8,205.9 | 2,819.3 | 3,163.5 | 1,928.8 | 294.3 |
| 2007 | 0.33 | 9,294.7 | 3,130.5 | 4,178.6 | 1,493.8 | 491.9 |
| 2008 | 0.39 | 9,825.4 | 2,880.7 | 4,589.9 | 1,783.6 | 571.2 |
| **Sichuan** | | | | | | |
| 1999 | 0.34 | 4,951.2 | 3,108.2 | 1,762.4 | 80.5 | 0.0 |
| 2000 | 0.34 | 5,580.3 | 3,217.5 | 2,111.0 | 251.8 | 0.0 |
| 2001 | 0.35 | 5,948.2 | 3,286.7 | 2,380.0 | 281.5 | 0.0 |
| 2002 | 0.33 | 6,591.0 | 3,439.6 | 2,747.9 | 403.5 | 0.0 |
| 2003 | 0.33 | 7,196.1 | 3,616.9 | 3,053.4 | 525.8 | 0.0 |
| 2004 | 0.33 | 7,709.0 | 3,881.8 | 3,261.4 | 565.8 | 0.0 |
| 2005 | 0.31 | 7,570.0 | 3,427.4 | 3,163.3 | 723.4 | 255.9 |
| 2006 | 0.31 | 8,540.3 | 3,847.2 | 3,651.6 | 767.0 | 274.4 |
| 2007 | 0.37 | 11,571.5 | 5,070.5 | 5,616.2 | 594.8 | 290.0 |
| 2008 | 0.39 | 12,445.6 | 5,316.8 | 6,157.8 | 554.4 | 416.6 |

Source: Yin and Liu (2011)

looked at the GfG's impact on inequality among 1,251–1,461 households (depending on the year) in six counties of Shaanxi and Sichuan provinces between 1999 and 2008. In 1999, the Gini coefficient for the households studied was 0.34.[13] The Gini coefficient dropped slightly until 2005–2006, then rose to be higher in 2008 than in 1999. Based only on these figures, we can say that the GfG ended up increasing inequality. Nevertheless, interesting information can be gathered about the sources of income, in particular the limited importance of GfG subsidies, and the increasing importance of off-farm work. In Shaanxi, off-farm incomes increased fourfold from Yuan 1,108.3 in 1999 to Yuan 4,589.9 in 2008, while in Sichuan it increased more than threefold from Yuan 1,762.4 in 1999 to Yuan 6,157.8 in 2008. Meanwhile, on-farm incomes remained almost unchanged in Shaanxi, while in Sichuan they increased by about 70 % (Table 11.15) (Yin and Liu 2011). Thus, the

[13] The lower the Gini coefficient, the more equality there is. A Gini coefficient of 0 means that everybody has exactly the same income. A Gini coefficient of 1 means that all income is concentrated in one person.

greatest change is in off-farm employment. Not unexpectedly, those willing and able (or forced) to leave the countryside, or engage in non-farm work in the rural areas, are able to earn much higher incomes than they did on the farm, which increases inequality. It is unknown whether the migrants are participants or non-participants, and other studies have found that they may be both. It is possible, however, that by encouraging off-farm employment of farmers with the worse land, the GfG may have contributed to the poorest farmers becoming the "new rich", thus reversing the social structure in the villages. Some of the poorest households may now be among the richest, thanks to off-farm work. These may be interesting consequences of the GfG that could be further investigated.

**Conclusions**

This chapter has looked at the changes in income levels among GfG participants, and compared the changes to those of non-participants. The conclusions are mixed. In some cases, non-participants have seen their incomes increase more than participants while in other cases the opposite is true. However, it is useful to remember that if the incomes of participants failed to increase (or even declined) it does not necessarily mean that the GfG failed: it is possible that participants' incomes would have declined even more without the GfG. In some cases, GfG subsidies constituted a relatively large part of participants' total income (e.g. Uchida et al. 2005), which means that a cut in subsidies will lead to a considerable drop in their standard of living. In other cases (e.g. Yin and Liu 2011), the income from off-farm work formed the largest component of total income. This is not entirely surprising, since off-farm wages are usually considerably higher than income from agriculture, especially when practiced on marginal land. Nevertheless, here too there are some contradictory findings among researchers. Some researchers (e.g. Liang et al. 2012) have found that participants have higher incomes, while others (e.g. Yao et al. 2010) have learned that non-participants have higher incomes from off-farm work. Thus, there are strong indications that the GfG has had different impacts in different areas, either because of differences in leadership (as pointed out by Yao et al. 2010), because of different local opportunities for off-farm work, or because of uneven environmental and ecological conditions.

# Chapter 12
# Labor Force Redistribution

**Abstract** This chapter reviews these issues in more detail, by looking at labor force redistribution. The Grain for Green was expected to increase migration and off-farm employment. However, researchers found that in some areas farmers who joined the Grain for Green program migrated less. These farmers were able to improve the productivity of their remaining fields (as well as farm fields of those who had left), which resulted in higher incomes and removed the need for migration. Successful migration also depends on qualifications and social networks, and it is likely that those who joined the Grain for Green, often the poorest members in the villages, had fewer opportunities to migrate that their wealthier neighbors. In any case, the evidence shows that there is considerable variation among communities in terms of the impact of the Grain for Green on income levels, and few generalizations can be made.

**Keywords** Labor market • Migration • Off-farm work • Agricultural fields' productivity • Income levels

## Introduction

One of the unstated goals of the Grain for Green program has been that of releasing labor from farming, and encouraging farmers to engage in alternative activities, either in their townships of residence or elsewhere. Indeed, the ultimate success of the GfG depends on its ability to restructure the employment opportunities of rural households so that farmers can raise the opportunity cost of their non-farm labor. This goal was facilitated by the dynamic economic growth that occurred in the 2000s, especially in the eastern provinces, which made it easier for farmers to find employment in labor-intensive industries. In the rural areas, encouraging out-migration reduced the oversupply of labor and increased the amount of land available to those who remained in the villages, making it possible to increase per-capita incomes. At the same time, encouraging farmers to seek off-farm employment increased the supply of labor in the eastern provinces, which helped keeping wages down and contributed to their sustained economic growth. This in turn helped the government raise additional funds from taxation that could be used to develop the rural areas. From this perspective, the GfG may

 
C.O. Delang, Z. Yuan, *China's Grain for Green Program*,
DOI 10.1007/978-3-319-11505-4_12

be said to have been part of a larger plan to transform the Chinese economy. This chapter describes the labor force redistribution generated by the GfG, both locally (off-farm employment in farmers' townships of residence), and through migration to local cities or the eastern provinces.

## Reallocation of Time Across Job Types in Villages

In this section we review some of the literature that looked at changes in labor allocation within rural villages. Zhou et al. (2007) estimated the changes in the amount of labor time expended for eight different types of work, after the introduction of the GfG. The research was carried out in Liping County (Guizhou province), and was based on survey data of 1,192 peasant laborers. Table 12.1 reports the increased and decreased allocation of labor among different activities due to the GfG program. The greatest increase in labor time was associated with "protecting forest" and "forest". Meanwhile, there was a slight decrease in agricultural work, with little change overall in the other activities, and a similar number of people spending less time than people spending more time, including, perhaps surprisingly, in "non-agricultural work".

Zhou et al. (2007) also looked at the changes in total income among the eight types of work (Column 2, Table 12.1). To do so, they used labor distribution ratios as the weight to provide the baseline information that can be used to compare the income generated by shifts in labor allocation among sectors. Zhou et al. (2007) concluded that even though labor reallocation caused by the GfG had positive

**Table 12.1** Percentage labor change after reforestation for each job type

| | | Labor time expended on different activities | | | | |
|---|---|---|---|---|---|---|
| Job Type | Income (US$/year) | Decrease a lot (%) | Decrease a little (%) | No change (%) | Increase a little (%) | Increase a lot (%) |
| Agriculture | 148.45 | 13.42 | 44.95 | 22.13 | 15.08 | 4.43 |
| Stock breeding | 100.81 | 6.22 | 9.27 | 37.34 | 37.07 | 10.10 |
| Cutting fuel wood | 42.65 | 28.77 | 29.32 | 32.23 | 8.58 | 1.11 |
| Protecting forest | 213.72 | 6.64 | 4.56 | 23.65 | 33.33 | 31.81 |
| Forest | 213.72 | 8.58 | 10.24 | 29.18 | 36.38 | 15.63 |
| Non-agricultural work | 367.8 | 5.39 | 7.61 | 62.79 | 15.91 | 0.83 |
| Foster children | 0 | 13.14 | 13.83 | 52.56 | 14.94 | 5.53 |
| Housework | 84.34 | 13.55 | 21.72 | 48.55 | 12.31 | 3.87 |
| Weighted mean income | 170.66 | | | | | |

Source: Zhou et al. (2007)

**Table 12.2** Change in labor allocation after reforestation

| | Big increase (%) | | Slight increase (%) | | No change | | Slight decrease (%) | | Big decrease (%) | |
|---|---|---|---|---|---|---|---|---|---|---|
| Job Type | M | F | M | F | M | F | M | F | M | F |
| Agriculture | 1.3 | 1.3 | 4.5 | 3.2 | 2.6 | 4.5 | 44.9 | 44.2 | 42.9 | 37.8 |
| Livestock production | 5.8 | 6.4 | 9.4 | 10.1 | 25.0 | 26.3 | 3.8 | 2.6 | 4.5 | 5.1 |
| Fuelwood collection | 0.0 | 0.0 | 1.9 | 1.3 | 26.9 | 24.4 | 3.8 | 4.5 | 7.1 | 5.1 |
| Forest protection | 70.5 | 62.8 | 14.7 | 16.0 | 0.6 | 2.6 | 7.7 | 5.8 | 0.0 | 0.6 |
| Forestry production | 1.9 | 1.9 | 5.8 | 6.4 | 11.5 | 10.3 | 0.0 | 0.0 | 0.0 | 0.0 |
| Non-agricultural work | 4.5 | 2.6 | 7.1 | 6.5 | 3.8 | 3.2 | 0.0 | 0.6 | 0.0 | 0.0 |
| Child nursing | 10.9 | 10.3 | 25.0 | 25.0 | 41.0 | 38.5 | 0.0 | 0.0 | 7.1 | 5.1 |
| Housework | 11.5 | 9.0 | 37.8 | 26.9 | 28.2 | 35.3 | 3.2 | 5.1 | 4.5 | 17.9 |

Source: Wang and Maclaren (2011)
Note: M and F represent male and female, respectively.

economic benefits, these benefits were not sufficiently large to motivate peasants' participation in the GfG without government subsides. The financial subsidies were essential to elevate peasants' incomes in poor regions such as Liping.

Wang and Maclaren (2011) carried out a similar study to that of Zhou et al. (2007) in Dunhua County (Jilin Province), and found similar changes in the allocation of time among eight types of work before and after conversion (Table 12.2). As in Liping County, the most significant increase in time allocated after the reforestation was for "forest protection", while the most significant drop was for "agriculture"; a majority (70.5 % of men and 62.8 % of women) claimed a "big increase" in time allocated to forest protection, and over 80 % of both men and women reported a "slight decrease" to a "big decrease" in time allocated to agricultural cultivation (Table 12.2).

Wang and Maclaren (2011) found that, apart from additional time spent caring for forests, the reduction in agricultural labor required after land conversion led to an increase in child care (21.2 % of men and 50 % of women declared a "big increase" or a "slight increase") and household work (49.3 % of men and 35.9 % of women declared a "big increase" or a "slight increase"). Wang and Maclaren (2011) also found that families with more members of working age had more surplus labor which increased the number of migrant workers. Of 156 respondents, 23 peasants (14.7 %) indicated that they had at least one family member who had become a migrant worker due to the land conversion project and that these workers were sending remittances home that averaged Yuan 7,500 per year. In terms of time allocated to livestock production and non-agricultural work, there was a slight increase after the land conversion (Wang and Maclaren 2011).

Xu et al. (2004) contended that the ultimate success of the GfG depends on its ability to restructure the production practices of rural households so that farmers can raise the opportunity cost of their non-farm labor (such as livestock production and off-farm employment). Indeed, Uchida et al. (2007) observed that, in many areas, the GfG was accompanied by efforts of villagers to increase their livestock enterprises. Such a shift, in fact, might have been expected since most of the compensation was paid in grain that could be used for feed. In addition, in some regions households that had more time available, especially after the first year of the program, were allowed to plant fodder, alfalfa, and other mulches that not only provided protection for the soil during the initial growth periods of the newly planted trees, but could also be used as feed (Uchida et al. 2007). The authors found evidence for a significant increase in livestock activity for program participants. In contrast, Cui (2009) suggested that the number of labor days spent on livestock activity has remained fairly stable in the Chang Ping district of Beijing in the post-GfG period, while the revenue from livestock activity was actually found to have declined after the GfG began.

## Changes in Labor Market and Off-farm Employment

Yin and Liu (2011) used a longitudinal dataset for the period 1999–2008 that contained a large number of households in six counties in Shaanxi and Sichuan Provinces, to examine labor allocation of GfG-participating households. The results are represented in Fig. 12.1. Both in Shaanxi and Sichuan provinces the share of on-farm work (in terms of person-days) decreased: by 23 % in Shaanxi and 28 % in Sichuan. Average household labor time in agriculture dropped from 227 person-days in 1999 to 175 person-days in 2008 in Shaanxi, and from 321

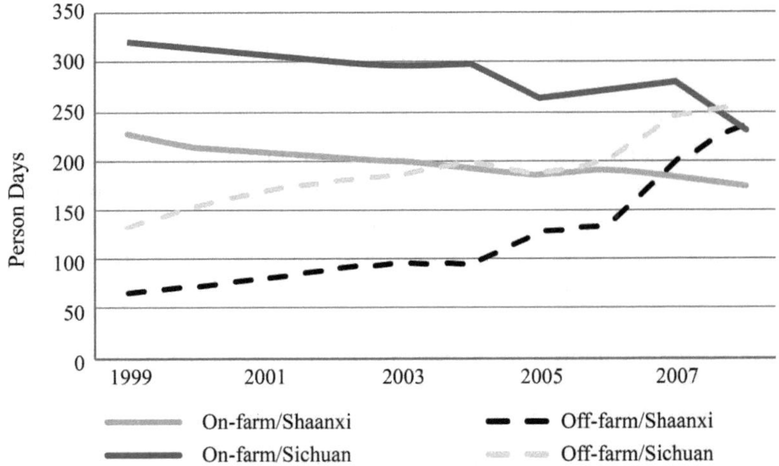


**Fig. 12.1** Labor allocation in agricultural and off-farm/off-village employment activities (Source: Yin and Liu 2011)

person-days in 1999 to 232 person-days in 2008 in Sichuan. On the other hand, off-farm work increased by 260 % in Shaanxi (from 66 to 238 person-days), and by 85 % in Sichuan (from 133 to 246 person-days). In comparison, non-participating households maintained a higher level of on-farm employment, while participating households sought higher levels of off-farm employment, either because they had the opportunity to do so, or because they were forced to do so. The extent to which the shift from on-farm to off-farm labor was voluntary is obviously questionable, but most researchers argue that it usually is. It is also striking to note that in the two provinces the total household labor, as measured in person-days, increased markedly over time. It seems that, as opportunities for employment increased from 1999 to 2008, people worked more. In Shaanxi, household members worked on average 293 person-days a year in 1999 and 413 person-days a year in 2008, an increase of 41 %. In Sichuan, they worked on average 454 person-days in 1999 and 478 person-days in 2008, an increase of 5 %.

A similar situation has been described by Wang et al. (2007b) for Yiyang County, Jiangxi Province. In particular, Wang et al. (2007b) found that the proportion of income earned by migrant workers increased from one-third of total income in 2000 to one-half in 2002.

Uchida et al. (2009) argued that household pre-program participation in off-farm labor markets may also be inhibited by low incomes, the absence of liquidity to finance the shift to the off-farm market, and poorly functioning land and credit markets (Hoff and Stiglitz 1990; Bardhan and Udry 1999). Indeed, high transaction costs, weak information-sharing, and other regulations have been shown to restrict farmers in rural China from starting small enterprises and seeking wage-earning jobs (deBrauw 2002; Knight and Song 2005). Furthermore, Uchida et al. (2009) pointed out that, although formal and informal loans are available, borrowing remains severely constrained, especially for the resource-poor strata of the population (International Fund for Agricultural Development 2001). Credit constraints have been shown to affect factor allocation in the production decisions of China's rural households (Feder et al. 1990). Since land rental markets are frequently incomplete in rural China, most households cannot leave agriculture entirely (Nyberg and Rozelle 1999). Given these conditions, if the GfG program can improve the liquidity of farmers, the program may enable them to find off-farm jobs and increase other productive activities (Uchida et al. 2009).

Drawing on household data collected from Gansu and other provinces, and descriptive statistics, Zhi (2004) shows that implementing the GfG has promoted the transfer of rural labor out of the farming sector. Hori and Kojima (2008) also found that, after the launch of the GfG program in Yulin district (Shaanxi province), the number of migrant workers among GfG program participants increased considerably.[1] Hori and Kojima (2008) found that the income from migrant workers accounted for more than 70 % of non-agricultural income for their households in

[1] Hori and Kojima's (2008) field research ran from July 29 to August 27, 2004, in Y village, Mizhi County, Yulin City, Shaanxi Province. The researchers also collected socio-economic data on 27 households, which were randomly selected from 226 households interviewed in Y village.

2004. Before the GfG began, manual labor was largely in mining and construction works. After implementation, the variety of migrant jobs diversified, although in 2004 more than 50 % of villagers were still working in the district of Yulin City. On the other hand, opportunities for off-farm work in the village were very limited: out of 27 households surveyed, three households run a small store, one household was a crop broker, and three households were cab drivers (Hori and Kojima 2008).

Other studies described similar situations. Peng et al. (2007) looked at the situation in Zhangye, a prefectural-level administrative area in Gansu Province, and found that the reduction in cropland resulted in a sharp increase in surplus labor. Most of the surplus laborers either migrated to other regions to work or engaged in non-agricultural work locally. To do this, they were helped by the local government. In the early 2000s, the governments in Zhangye helped a large number of local laborers, including those released by the GfG project, to find jobs in other regions. According to interviews with local officials, the prefectural government helped the peasants not only by providing information about job opportunities available in large cities, but also by improving their working skills through various professional training activities. In Zhangye, labor migration has proved to be an important measure to increase local rural household incomes (Peng et al. 2007).

Similar arguments have been made by Wang and Maclaren (2012), who looked at the case of Dunhua City, Jilin Province, and argued that the Korean minority welcomed the GfG enthusiastically, as it helped them move out of labor-intensive farming activities and migrate to South Korea. As in the case of Zhangye described by Peng et al. (2007), this migration was organized by local governments. Ge et al. (2006) found similar developments in Wuqi County, Shaanxi Province. After 103,700 ha of their cropland were reforested, 15,000 farmers switched from farming to construction, transportation, and restaurant businesses between 1998 and 2003.

Uchida et al. (2004) also argued that in addition to forest incomes, off-farm labor is considered another important alternative source of income for participant households, once their labor is released from cultivation. However, according to Liang et al. (2012) many rural households are exposed to market imperfections and the lack of institutional infrastructure, such as local schools, means that they are still not able to efficiently reallocate their labor to other activities, particularly in cities.

Liang et al. (2012) discussed another impediment to migration. Through a survey in Zhouzhi County, Shaanxi province, they found that migration was related to the composition of and number of children in a household, rather than participation in the GfG. Households with children but without elderly tended to have lower migration rates (and lower incomes after participation in the program). Since they did not have elderly members able to care for the children, the adults of working age chose not to migrate.

Similar conclusions were reached by Uchida et al. (2007), who found no evidence of significant program effects on the employment prospects of adults in the household, in any of the three approaches they used (propensity score matching [PSM], difference-in-differences [DD], and difference-in-differences matching [DDM]) (Table 12.3). Indeed, households were likely to use the newly freed-up

**Table 12.3** Estimated effects of the GFG on changes in employment prospects using three approaches, 1999–2002

| | | *PSM* | *DD* | *DDM* |
|---|---|---|---|---|
| | Dependent variable | Y(2002) | Y(2002)–Y(1999) | Y(2002)–Y(1999) |
| Diversification of economy | Non-agricultural income per capita (Yuan) | −80.38 (0.82) | −2.92 | −29.28 |
| | Off-farm work (number of adults with off-farm work in household) | −0.04 | 0.0034 | 0.045* |
| Labor redistribution | Migration status (number of adult migrants in household) | −0.04 | −0.018 | −0.04 |

Source: Uchida et al. (2007)
Notes: (1) The estimates are adjusted for inflation
(2) * significant at 10 %

**Table 12.4** Estimated effects of the GfG on changes in value of house and other major assets using three approaches, 1999–2002

| | *PSM* | *DD* | *DDM* |
|---|---|---|---|
| Dependent variable | Y(2002) | Y(2002)–Y(1999) | Y(2002)–Y(1999) |
| Value of house (Yuan) | 485.80** | 323.13 | 521.80** |
| Fixed productive assets (Yuan) | 320.23 (0.89) | 682.97 (1.36) | 312.77 (0.80) |
| Livestock inventories (Yuan) | 180.00 (2.94)*** | 161.15 (1.46) | 220.02 (2.74)*** |

Source: Uchida et al. (2007)
Notes: (1) The estimates are adjusted for inflation
(2) ** significant at 5 %
(3) *** significant at 1 %

labor in the first years of the program, not for activities that led immediately to greater incomes, but for investing in other parts of their asset portfolios, including home construction. According to Uchida et al. (2007), many farmers reported that the reduced need to spend time in agriculture due to the lower labor requirement of the newly planted forests after the first year, allowed them to spend their time building a new house or renovating their old one. In rural China housing is by far the largest asset in a household's portfolio and many households spend hundreds of hours building their houses. This is especially true in some forested areas where the GfG program is implemented, since households use their labor to cut trees and saw them into planks that can be used in home construction. Indeed, by 2002 the value of the houses of participating households increased significantly (Table 12.4) (Uchida et al. 2007).

Some analysts have also questioned the GfG program's effects on labor transfer. An important question is whether the GfG had a significant impact in reforming the local economies and encouraging off-farm work and migration to cities, or whether those changes had already begun and would have continued regardless of the

GfG. Uchida et al. (2005) addressed this issue by comparing the situation in 1995, before the start of the GfG, to that in 1999 and 2000, shortly after its implementation in Ningxia Province and Guizhou Province. Their study showed that by 1995, the economy in these provinces was already undergoing changes and, if anything, the GfG slowed down the process. The number of households gaining incomes from off-farm work increased by 3 % a year from 1995 to 1999, but only by 2 % from 1999 to 2000 for Ningxia Province. For Guizhou Province the data are even more striking: the number of farmers engaging in off-farm work increased by 8.5 % a year from 1995 to 1999 but did not change from 1999 to 2000 (Fig. 12.2). The other income generating options described in Fig. 12.2 are similarly inconclusive as to the GfG's impact on economic diversification of the rural economy. These results have to be weighted against two factors. First, the GfG is unlikely to have made a

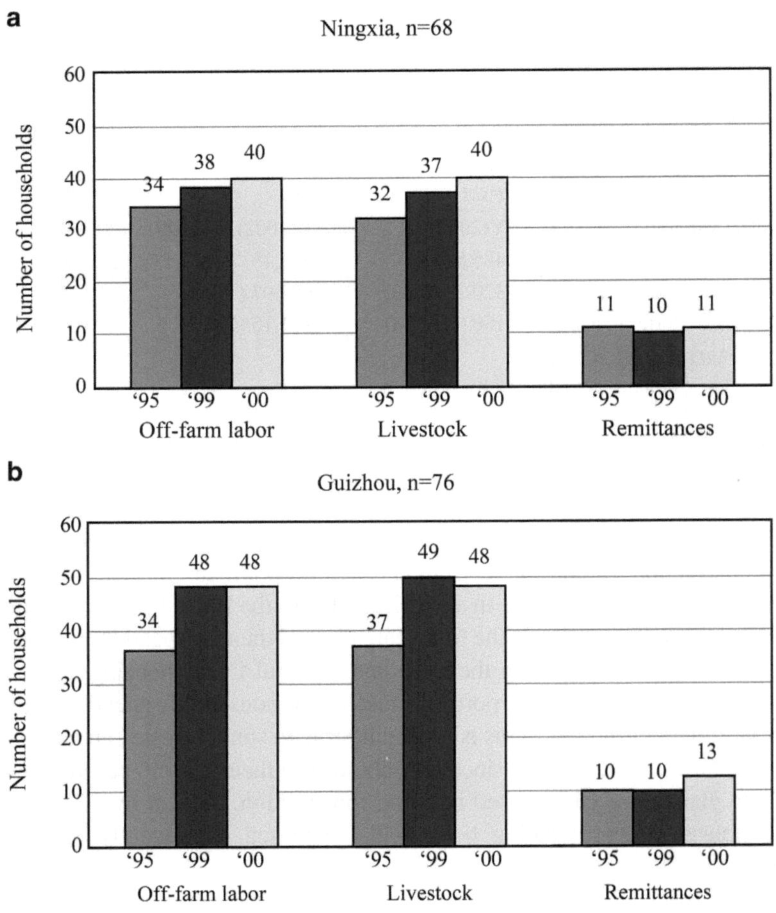


**Fig. 12.2** Change in the number of households participating in GfG Program with Income from Off-farm Labor/Businesses, Livestock and Remittances in Ningxia and Guizhou, 1995–2000 (Source: Uchida et al. 2005)

considerable impact in its first year. Second, the sample is restricted to participating households, and lack of comparison with non-participating households makes it difficult to understand the real impact of the program (Uchida et al. 2005).

Like Uchida et al. (2005), Xu et al. (2004) found that in the areas they studied in Shaanxi, Gansu, and Sichuan Provinces, the GfG did not have a significant impact on household labor allocation in non-farming sectors until 2002. Similarly, Liang et al. (2012) suggested that the increase in off-farm work may simply reflect the long-term trend toward greater reliance on off-farm work by farm households, or risk-coping strategies, rather than being a result of the GfG reforms.

Uchida et al. (2007) argued that the evidence that participating households had begun to shift their labor into the off-farm sectors was weak. Instead, it was often an ex-post facto proposition: for households that had already sent a household member into the migrant labor force before the GfG, the program may have become an opportunity to take some cropland out of production. This could be attractive because, compared to cultivation, many forestry activities require less labor (Uchida et al. 2007). Interviews before and during the implementation of the GfG frequently found that local leaders and villagers expected that setting aside an appreciable amount of their land would shift labor into the off-farm sector. In many cases, however, it seems that such expectations were not realized, since finding an off-farm job may require more than a desire to work off the farm (Uchida et al. 2007). Connections, human capital, and social capital are also necessary to access off-farm jobs, and many rural households do not possess these determinants. Furthermore, since Uchida et al. (2007) only looked at labor allocation decisions during the first 3 years after the start of the GfG, it may have been too soon to detect changes.

Based on longer longitudinal data, Uchida et al. (2009) consistently found that, on average, the GfG had a positive (although only moderate) effect on off-farm labor participation; households that participated in the program were increasingly allocating their family's labor to the off-farm labor market. The results also indicated that households with less liquidity before participating in the program were more likely to start off-farm jobs, supporting the view that the compensation for setting aside cultivated land may be relaxing liquidity constraints, allowing participants to more easily move into the off-farm sector relative to non-participants. Yet, the extent to which off-farm work may be sought is also conditional on an individual's skills, s age and education, and is more popular among younger people who have achieved a higher level of education (Uchida et al. 2009).

Yao et al. (2010) attributed the conflicting findings of different researchers to the program's broad coverage and the varying biophysical and socioeconomic conditions across the country. They pointed out that it seems unrealistic to expect a uniform outcome for such a large program. Furthermore, several factors could have affected the studies' findings: the location of the sites that were selected, the kinds of data that were collected, the year the study was carried out, and the length of the fieldwork period. In addition to household characteristics, the level of local economic development, the number of years the program was implemented, and local political leadership, could all influence the impact of the GfG on labor transfer. Yao et al. (2010) tested these propositions in three adjacent counties – Wuqi, Dingbian, and Huachi – in the Loess Plateau region, covering periods before and after the program's inception (1999 and 2006). These counties belong to different jurisdictions, which can better reflect the varying extents of program execution, political setting, and economic development. The results of off-farm employment and income regressions highlight the fact that participating in the program has accelerated the transfer of farming labor and has greatly stimulated income growth from off-farm opportunities.

Cui (2009) also discussed differences among GfG participants in terms of off-farm labor allocation, but this time based on the amount of capital available to the households. Combining a difference-in-difference (DD) estimator and a switching regression model with unobserved sample separation, they found that the impact of the GfG on off-farm labor allocation is negative for capital-unconstrained households, while for capital constrained households it is positive and significant. This indicates that participants in the GfG have increased the off-farm labor supply of constrained, and presumably poor, households (Cui 2009).

**Conclusions**

This chapter has looked at labor force redistribution that followed the implementation of the GfG. Researchers tend to agree that the GfG has generated a number of surplus laborers in the rural villages; however, not all agree on what happened to this surplus labor, probably because the variety of local conditions led to a diversity of outcomes. In some areas, the GfG simply shifted labor allocation locally, from the farm to the forest or to animal husbandry, or in some cases to the expansion and modernization of the house, as people were free to engage in activities other than farming. In other cases, the GfG resulted in migration to seek off-farm jobs, contributing to and facilitating a surge in migrant labor across China. For example, in Guizhou Province, the number of migrant workers increased 48 % (from 2.2 to 3.1 million) between 2000 (before the GfG) and 2005 (Yang 2006). Indeed, Xu et al. (2007) argue that the GfG program contributed to the social transformation of traditional rural society by enabling the workers liberated from on-farm work to seek off-farm jobs in or beyond their locale. Some researchers argued, however, that this migration was already under way before the implementation of the GfG, and that those who migrated were often better off farmers with a good social network, who were not necessarily the people targeted by the GfG. Indeed, successful migration depends not only on capital (which the GfG provided), but also on marketable skills and education levels, which the GfG did not always provide – although other programs, as well as local authorities, helped in some cases. Liang et al. (2012) also pointed out that other characteristics of households, such as household composition and the availability of elderly members able to care for children, also contributed to determining whether household members would migrate.

# Chapter 13
# Sustainability of the Grain for Green Program

**Abstract** In this chapter, we examine the sustainability of the Grain for Green. The Grain for Green was originally scheduled to end in 2007, but was extended, and the subsidies are now set to end beginning in 2015 for the land that was first set aside, and later for other land. The question is whether the farmers will continue with the Grain for Green-induced land use changes, or will revert back to the pre-Grain for Green land uses once the subsidies end. There are constraints on cutting the trees, in particular a quota system, whereby the farmers need to obtain permission from the Forestry Bureau to fell their trees. Nevertheless, if the income from tree products do not compare favorably to those from cash crops, when the subsidies end there will be considerable pressures on the forest. The hope is that the rural economies have sufficiently transformed (through the Grain for Green and other programs), and that off-farm opportunities abound, so that farmers no longer need to revert their least productive land to pre-Grain for Green land uses. One issue that complicates assessments of sustainability is the fact that most studies were done during the first years after the program was implemented, when the monetary benefits from the economic trees could not yet be fully ascertained. However, some studies did try to estimate future changes in prices, and predict farmers' adaptation to such changes.

**Keywords** Future incomes • Changes in taxation levels • Changes in interest rates • Farmers' attitudes • Property rights • Program sustainability

## Introduction

At the time of implementation, government financial transfers made the Grain for Green program very profitable for farmers because, in many cases, subsidies were higher than incomes from farming the set-aside land. Subsidies were essential if farmers were to join the program voluntarily, since during the first few years farmers could make little money from the sale of tree products (such as fruits) or from thinning the trees. However, the GfG, like many other reforestation and rural development programs in the developing world, has a limited budget and a finite time line, and subsidies will eventually end. The GfG is very costly, with a projected investment of no less than Yuan 431.8 billion by 2016. Originally, it was set to last for 8 years for ecological trees, 5 years for economic trees, and 2 years for



C.O. Delang, Z. Yuan, *China's Grain for Green Program*,
DOI 10.1007/978-3-319-11505-4_13

grassland, with the program beginning to be phased out beginning in 2007. In 2007, however, the government renewed the GfG program payments and the program is now scheduled to begin being phased-out in 2015. The GfG has been renewed once, but it is unclear whether the government will continue program payments after this second period ends. What is clear is that subsidies cannot continue forever, and that if deforestation resumes once the payments end, the reforestation program can be said to have failed. With every reforestation program, however, there is the question of whether it is sustainable (i.e. whether the land use/land cover introduced by the program will continue after payments end).

Evidence from similar land set-aside programs in other parts of the world suggests that once payments cease, a large amount of land may return to its pre-program use (Uchida et al. 2005). This is the case, for example, of the Conservation Reserve Program (CRP) in the United States, where a relatively high number of farmers revert their land back to cultivation after subsidies end (USDA 2000). The objectives of the CRP do not include poverty alleviation, and the CRP is not intended to lead to economic diversification of the targeted areas. Since the GfG includes these objectives, it is hoped that it will be more sustainable than the CRP. Available evidence suggests, however, that officials in China should be concerned about the long-term sustainability of the GfG.

Farmers may chose not revert the land to pre-GfG land uses for three reasons. First, they may earn more from the economic or ecological trees than they would earn from growing food crops. Second, the GfG (together with other programs undertaken by the Chinese government) may has been successful in transforming the local economy, and there are now better (in farmers' views) off-farm opportunities locally. Third, they may have migrated to cities, either in their province of origin or in other provinces, with better opportunities than their place of origin, and have now settled in those cities. Thus, it is fair to say that the sustainability of the GfG also depends on the broader economic development of China, and on how successfully the farmers can diversify their livelihood (Hori and Kojima 2008).

This chapter looks at how sustainable the GfG can be expected to be, that is, whether farmers are likely to revert the converted land back to pre-GfG land uses once the subsidies end, or whether the conditions have been put in place for GfG-land uses to be maintained, not only because of legal restrictions on tree felling,[1] but also because the farmers chose to continue the land-use practices introduced by the GfG. We first examine the economic benefits from plants and how these compare to food and cash crops, given expected (or potential) changes in prices, taxation, subsidies, and interest rates. This addresses the question of how well future expected profits from tree products compare to profits from pre-GfG land use. By the time the subsidies end, the land use/land cover changes introduced through the GfG should ideally generate an income that is comparable (or superior) to alternative, more

[1]As mentioned in Chap. 1, there is a quota system for felling trees. If farmers cut trees without the necessary permit, they have to pay a fine or face imprisonment. Also, in most rural villages there are other programs (Zhang et al. 2006), and if people illegally fell trees the subsidies they receive from other programs may also end.

destructive, uses of the land, which existed before GfG conversion. If higher incomes do not materialize, there is a risk that farmers still living in rural areas will convert the land to pre-GfG land uses. We then turn to farmers' attitudes. These are important, because economic issues are only some of the factors farmers take into consideration when deciding which crops to grow or which ventures to undertake. Finally, we look at property rights of land, plants, and forest products. Researchers have found that considerations of property rights play an important role in farmers' decisions regarding what to do with the land.

## Present and Future Income from Plants

Some researchers have expressed doubts about the stability of GfG-related incomes. For example, Xu et al. (2010) argues that the future value and shorter-term income-generating capacity from ecological trees and economic trees (orchards) planted under the GfG do not look promising. For ecological forests, this is due to low survival and slow growth rates of ecological trees in many regions, as a result of low rainfall levels and unsuitable abiotic conditions for timber trees (especially in the arid northwest provinces of Gansu and Shaanxi), uncertainties about the future of China's forest sector reforms, and the potential oversupply of timber due to large-scale plantations in the south. For economic forests, the fast expansion of the GfG has led to many different regions in China planting similar orchard crops, raising concerns about future oversupply and price stability (Xu et al. 2010).

However, other researchers have argued that the incomes from economic trees should compare positively to those of food and cash crops. Xie et al. (2006) looked at the opportunity cost of the converted land, and how it compared to the profits that can be generated post-conversion (the analysis is on 20 to 30 households that participated in the GfG in four counties in Qinghai and Shaanxi Provinces). The authors provided output and prices forecasts until 2018 for agricultural food crops, cash crops, and timber products (at different discount rates and output prices), to simulate the future incomes from different land uses. They found that the potential revenues from converted land are relatively attractive, leaving little concern about the sustainability of the program after the government subsidies expire.

Zhou et al. (2007) addressed the same issue, setting out to compare the economic returns (per hectare and labor input) of different tree species to those of food and cash crops in Liping County, Guizhou Province. Liping County is 4,441 $km^2$ large, with an elevation that ranges from 600 to 1,500 m. Its landscape is dominated by mountains and hills. The physical properties of the county, as well as its isolation, contribute to the poverty of the county; peasant net income per capita was approximately US$153, lower than the national average of US$183.13 (Tang 2000).[2] Liping's economy is closely tied to its agricultural and forestry production. The

[2] The county's population in the year 2000 was 489,000, with 82 % of the population minorities. Administratively, there are 25 towns/townships and 403 villages. In 2001 there were 244 villages

implementation of the GfG program led to a decrease in both crop yields and timber output, thereby reducing peasant household income. For this reason, financial subsidies were a critical factor for the successful implementation of the GfG program.

Zhou et al. (2007) looked at two issues. First, they compared the incomes of GfG-introduced trees and traditional food and cash crops, then examined the importance of subsidies for the successful implementation of the GfG. Second, they looked at the long-term potential incomes from these trees, and how well the incomes will compare to food and cash crops once their full potential profits are generated. Their analysis was based on two peasant household surveys conducted in 2003 and 2004.[3]

With the first issue, Zhou et al. (2007) confirmed the importance of subsidies to alleviate the losses that peasants incur when replacing food and cash crops with economic or ecological trees. Common food and cash crops planted on the slope land in Liping County are sweet potatoes and potatoes, which generate a net income of US$247.70 per hectare per year. On the other hand, trees do not produce any income during the first years, while planting trees involves higher costs during the seeding and planting stage. The study reported that, given tree plantation costs during the first year, the net land productivity and the net labor productivity was negative for most tree species, except for orange and bamboo plantation.

Government subsidies were US$415.60 per hectare per year, much higher than the opportunity cost of planting crops on sloping agricultural land (Zhou et al. 2007). Without the financial subsidies, the peasants would have lost money in carrying out the conversion of agricultural land to forest land. However, when taking the program subsidies into account, the economic situation of the surveyed peasants was drastically different, with the net economic return per hectare becoming positive for every tree species planted (Zhou et al. 2007). With the subsidies, the net economic return of the slope land, which is often low-quality marginal agricultural land, reached as much as US$588 per ha for orange trees, US$513 per ha for bamboo, and US$503 per ha for oil tea seed. With the financial subsidies, the area-weighted net economic return of land use for all tree species was US$385 per ha, which was higher than the value of grain production (US$326 per ha) and cash crops production (US$288 per ha). Considering income per person-day, with government subsidies, the highest net labor return was US$6.50 per day for Masson pine, which can be compared to a return of US$0.82 per day for sweet potatoes, and US$0.90 per day for rice. Masson pine and oil tea seed plantations do not require large labor inputs and therefore enjoy high gross labor productivity (Zhou et al. 2007).

with a total population of 226,000, in which peasant annual net income per capita was below US$145 (Zhou et al. 2007).

[3] The social and economic data of for reforestation in Liping County are derived from statistical yearbooks from 1999 to 2004, development reports by Liping authorities, and publications on local agriculture and forestry. Zhou et al. (2007) conducted interviews with government officials in the Liping Forestry Bureau, other agencies, and farmers. The researchers interviewed 471 peasant households from 21 towns and 76 villages; survey information covered 1,192 peasants. The total reforested area of respondents was 629 ha, equivalent to 7.6 % of the total reforested area (4,334 ha) in Liping. Minorities accounted for 71.3 % of the 1,192 peasants sampled (Zhou et al. 2007).

Zhou et al. (2007) argued that the Liping case illustrates the importance of government financial subsidies. These subsidies have been essential in making the project economically feasible for peasants because, in the short run, before the revenue from economic and ecological trees are fully realized, net revenue generated from the tree plantation is lower than that from producing agricultural products. Meanwhile, since subsidies were also higher than the incomes from food crops, they were a major means of elevating farmers' income.

The second issue addressed by Zhou et al. (2007) is that of the economic returns of plantations once the trees provide their full economic potential. To do so, they calculated the average yearly revenue of planted forests with perpetual rotation using the following equation:

$$\bar{R} = \frac{1}{T}\sum_{t=0}^{T} R_t e^{-rt}$$

where $\bar{R}$ is the average annual revenue during a rotation period, $R_\tau$ is the revenue per hectare at time $\boldsymbol{t}$, $\boldsymbol{e}$ is the base of natural logarithms, $\boldsymbol{r}$ is the discount rate (Zhou et al. (2007) use a discount rate of 5 % based on Alig et al. (1997)), and $\boldsymbol{T}$ the rotation period (Zhou et al. (2007) use Pan et al. (2004), to estimate the rotation period of the tree species in the surveyed area). To calculate the yield per hectare and the unit price of forest products (such as lumber and fruits) during each rotation period, Zhou et al. (2007) used empirical estimated data provided by the local officials from the Liping forestry sectors.

They concluded that among all tree species, tea plantation will potentially provide the highest economic return of US$3,565 per ha per year (US$3,666 per ha with the subsidy for 5 years). This was more than ten times the income from rice. Other trees, such as chestnut, pear, and orange also had great potential for economic rewards. On the other hand, the economic returns of Sawtooth Oak and Oil Teaseed were found to be very low, and were not economically viable options (Table 13.1). Zhou et al. (2007) also calculated the area-weighted annual average potential net income for the sampled area. The calculated value was US$661 per ha (US$778 per ha with the subsidy for 5 years), as compared to US$385 per ha for 2003–2004, under the conditions prevalent at the time of the fieldwork. Zhou et al. (2007) concluded that the economic prospects of tree plantation over the long-term were expected to be much better than the short-term economic benefits. Hence, if the early 2000s market conditions hold, tree plantation through the GFG project will provide substantially higher incomes to Liping's peasants than food production (Zhou et al. 2007).

Future incomes will obviously be determined by changes in the prices of the products. However, there may also be other changes that will affect future incomes, in particular taxation levels and interest rates. This issue has been addressed by Liao and Zhang (2008), who carried out research among 40 randomly selected households in Zigui county (Hubei province) in 2000–2001.[4] With the help of a questionnaire,

[4] Liao and Zhang (2008) choose Zigui County, in the Three Gorges Region in the western Hubei province, as their study area because it is representative not only of Hubei province, but also of the

**Table 13.1** Potential annual net income of trees in sample areas

| Species | Area (ha) | Rotation Period (year) | Total year of subsidy (year) | Potential annual net income ($ per ha) | Potential annual net income with subsidies |
|---|---|---|---|---|---|
| Tea | 7.77 | 25 | 5 | 3,565.03 | 3,666.09 |
| Chestnut | 1.78 | 25 | 5 | 1,719.81 | 1,784.33 |
| Tuliptree and hackberry | 4.69 | 21 | 5 | 1,279.89 | 1400.20 |
| Pear | 8.63 | 25 | 5 | 828.68 | 893.20 |
| Masson Pine | 51.37 | 25 | 8 | 752.57 | 853.63 |
| Orange | 5.19 | 25 | 5 | 678.79 | 743.31 |
| Chinese fir | 66.68 | 25 | 8 | 439.78 | 540.84 |
| Wild pepper | 0.82 | 25 | 5 | 473.02 | 537.54 |
| Bamboo | 31.31 | 11 | 8 | 255.26 | 484.94 |
| Sawtooth oak | 3.68 | 5 | 8 | 78.56 | 96.63 |
| Oil teaseed | 5.91 | 25 | 5 | 1.53 | 66.05 |

Source: Zhou et al. (2007)

Liao and Zhang (2008) asked about input costs, yield benefits, management regimes for five types of land use options, and characteristics of farmers and their participation in the GfG program. The land expectation value (LEV) method was used to examine the allocation of forest land among alternative options, based on the assumption of perpetual land use. LEV is estimate from the Faustmann model, a standard economic model to estimate land expectation values in forestry (Liao and Zhang 2008). The modified formula is

$$\mathbf{LEV} = \frac{\sum_{t=0}^{T}(R_\tau - C_\tau)\times(\mathbf{1}+\mathbf{r})^{T-\tau}}{(\mathbf{1}+r)^T - \mathbf{1}}$$

where $R_\tau$ denotes the revenue in the year $t$; $C_\tau$ stands for the cost in the year $t$ (including establishment cost $C_0$); $T$ is the rotation age; and $r$ is the interest rate.

The LEV assesses the gain or loss of shifting the farm lands to other land uses with changing interest rates, prices, wage rates, and tax rates.

upper reaches of the Yangtze river in terms of ecology, geographic factors, socio-economic conditions, and the significant number of orchard trees (specially citrus and chestnut), tea and pine plantations growing there. First, four villages were randomly drawn from a list of villages at the Forestry Administration in the county. Then from each village, ten households were randomly drawn from the village (Liao and Zhang 2008).

**Table 13.2** Comparison LEV for five types of land use options with tax or without tax at different interest levels

| Percent interest rate | Land use options | With tax | | Without tax | | Tax reduction only for pine tree by 50 % | |
|---|---|---|---|---|---|---|---|
| | | LEV (Yuan/ha) | Optimal rotation (year) | LEV (Yuan/ha) | Optimal rotation (year) | LEV (Yuan/ha) | Optimal rotation (year) |
| 4 | Crops | 26,396 | 1 | 47,480 | 1 | 26,396 | 1 |
| | Pine tree | 31,097 | 29 | 41,208 | 28 | 36,153 | 28 |
| | Citrus | 86,469 | 25 | 119,907 | 25 | 86,469 | 25 |
| | Tea | 67,134 | 28 | 128,080 | 28 | 67,134 | 28 |
| | Chestnut | 107,404 | 29 | 136,688 | 29 | 107,404 | 29 |
| 8 | Crops | 13,198 | 1 | 23,740 | 1 | 13,198 | 1 |
| | Pine tree | 7,632 | 23 | 10,548 | 22 | 9,089 | 22 |
| | Citrus | 24,213 | 27 | 38,436 | 27 | 24,213 | 27 |
| | Tea | 15,678 | 30 | 43,597 | 31 | 15,678 | 30 |
| | Chestnut | 40,880 | 31 | 53,589 | 30 | 40,880 | 31 |
| 12 | Crops | 8,799 | 1 | 15,827 | 1 | 8,799 | 1 |
| | Pine tree | 2,133 | 19 | 3,358 | 19 | 2,745 | 19 |
| | Citrus | 3,871 | 29 | 11,783 | 29 | 3,871 | 29 |
| | Tea | −1,687 | 30 | 15,263 | 32 | −1,687 | 30 |
| | Chestnut | 18,741 | 33 | 25,950 | 31 | 18,741 | 33 |

Source: Liao and Zhang (2008)

Note: The optimum rotation age is when the marginal value of holding the current stand is equal to the marginal cost of the land for renting plus the foregone interest payment for timber growth

## *Changes in Taxation Levels*

Removing the tax would increase incomes from trees compared to food crops, especially with an interest rate of 8 % (as it was during the time of the fieldwork) or higher (Table 13.2). Nevertheless, pine trees would still generate a lower income than food crops at 2000–2001 price levels. Even with a reduction in tax by 50 % for pine trees, the trees would be able to compete with food crops only if the interest rate dropped to 4 %. When interest rates are higher, removing taxes has the greatest positive impact in generating higher incomes from trees instead of food crops.

## *Changes in Interest Rates*

Liao and Zhang (2008) found that different land use options respond differently to interest rate changes (Table 13.2). All five land use options have greater LEVs with an interest rate of 4 %, followed by 8 %, and then 12 % (12 % giving the lowest LEV). Orchard trees and tea are more sensitive to interest rate changes than crops and pine trees. The possible explanation is that more investment is needed to establish orchards at the beginning of the production cycle. Pine plantations are less sensitive to interest rate changes, but shifting farmland to pine plantations can generate benefits only if the interest rate is low (4 %), possibly because the price of pine timber during the fieldwork period was low (350 Yuan/m$^3$ on average). Similarly, citrus and tea generate lower profits than crops with an interest rate of 12 %, but higher profits at lower interest rates. Compared to crops, only chestnuts have no economic loss, regardless of interest rate and taxation rate. In contrast to trees, crops are not as sensitive to interest rates because, unlike orchard trees and tea, crops do not require great investment at the beginning of the production cycle (Liao and Zhang 2008).

The data strongly suggest that credit markets are very important to farmers. If low interest rate loans are available, the financial returns of orchard trees are higher crops, even without government subsidies (Liao and Zhang 2008). This means that farmers might be willing to convert their farm land to orchard trees and tea without subsidies. The government could cut subsidies for orchard trees and tea, and farmers would still find them more profitable to grow than food crops. Since GfG subsidies cannot last forever, Liao and Zhang (2008) concluded that market-based approaches, such as developing credit markets and lowering interest rates for farmers, could facilitate implementation of the GfG and reduce its costs.

## *Subsidies*

The provision of government subsidies is sufficient to motivate farmers to shift their farm lands to other uses. Table 13.3 demonstrates that the government-initiated subsidy program facilitates shifts in land cover. When subsidies for citrus, tea, chestnut, and pine are delivered to farmers for 5 years, all four land use options generate higher land values than crops, no matter how much the interest rate changes. Under these circumstances, farmers who are land value maximizers could be willing to shift their agricultural lands to planting pine, orchard trees and tea (Liao and Zhang 2008).

Overall, over 90 % of farmers who were actively involved in the GfG program were satisfied with the program and were willing to shift their farms to forest lands. Still, farmers preferred orchard trees and tea to pine trees because the former generated higher returns than pine trees with the same subsidies (Liao and Zhang 2008). For these reasons, it would be efficient for the government to cut subsidies for economic trees and use the savings to increase subsidies for ecological trees, matching the subsidies to the economic benefits that can be obtained from each species.

Liao and Zhang (2008) suggested that, if the government carried out a cost-benefit analysis of different land uses, including the environmental benefits generated from land conversion, it would be able to determine which land use option was best for each region. Moreover, the authors argue that multiple incentive programs should be developed jointly. For example, whereas the agricultural tax in China has

**Table 13.3** Comparison of LEV for five types of land use options, with or without subsidy at different interest levels

| | | Without subsidy | | With subsidy | |
|---|---|---|---|---|---|
| Percent interest rate | Land use option | LEV (Yuan/ha) | Optimal rotation (year) | LEV (Yuan/ha) | Optimal rotation (year) |
| 4 | Crops | 26,396 | 1 | 26,396 | 1 |
| | Pine tree | 31,097 | 29 | 60,142 | 24 |
| | Citrus | 86,469 | 25 | 111,508 | 24 |
| | Tea | 67,134 | 28 | 89,742 | 26 |
| | Chestnut | 107,404 | 29 | 130,478 | 27 |
| 8 | Crops | 13,198 | 1 | 13,198 | 1 |
| | Pine tree | 7,632 | 23 | 26,549 | 18 |
| | Citrus | 24,213 | 27 | 40,161 | 25 |
| | Tea | 15,678 | 30 | 30,467 | 29 |
| | Chestnut | 40,880 | 31 | 56,256 | 29 |
| 12 | Crops | 8,799 | 1 | 8,799 | 1 |
| | Pine tree | 2,133 | 19 | 16,989 | 18 |
| | Citrus | 3,871 | 29 | 16,856 | 27 |
| | Tea | −1,687 | 30 | 10,750 | 30 |
| | Chestnut | 18,741 | 33 | 31,554 | 30 |

Source: Liao and Zhang (2008)
Note: Subsidy for pine trees and cash trees for 5 years (Yuan 3,450/ha per year)

been cut gradually, the timber tax is still high. If this tax was cut to the same level as the tax rate for agricultural crops (10 %), the LEV of pine plantations could catch up with that of crops, since prices of pine timber will probably increase by 30 %, given the implementation of the Natural Forest Protection Program (NFPP) since 1998 (Liao and Zhang 2008).

## Farmers' Attitudes

Regardless of the actual profits that farmers may make from the production of trees, tree products or crops, farmers' attitudes are equally important. Farmers not only look at total profits or price stability, but also consider a range of non-economic issues and may choose to continue growing trees even if the production of food or cash crops would be more profitable. Shi and Wang (2011) looked at farmers' attitudes towards the GfG adapting Bossel's orientation theory (1999). Shi and Wang (2011) designed seven orientors (Table 13.4), and collected data by asking farmers in Mizhi County, Shaanxi province yes-or-no questions. For each measure, the coordination coefficient U was calculated based on the number of "yes". The higher degree of U, the more positive the farmers' replies. "Security" (measured by farmers' net income) scores the lowest, which indicates that many farmers believe the

**Table 13.4** Social coordination coefficients of the indicators in Mizhi County

| Orientor | Question | U |
|---|---|---|
| Existence | "Does the GfG project affect the grain supply to your family?" | 0.75 |
| Project efficiency | "Have you converted all 25-degree-and-over sloped farmland to forest?" | 0.99 |
| Living choice | "Will you support the GfG project when food subsidy is cut off at the project end?" | 0.85 |
| Security | "Does the GfG project increase your net income?" | 0.70 |
| Adaptability | "Does grain subsidy make up for your loss in the GfG project?" | 0.85 |
| Coexistence | "Does the GfG project enhance your environmental consciousness?" | 0.99 |
| Psychological satisfaction | "Are you satisfied with the vegetation coverage after the implementation of the GfG project?" | 0.96 |
| Coordination Degree | | 0.87 |

Source: Shi and Wang (2011)

GfG did not increase their net income (Table 13.4). Similarly, many farmers found that their households' grain supply was not affected by the GfG (as measured by the indicator "Existence", whose U is 0.75). The other orientors' U range from 0.85 to 0.99, indicating farmers' positive attitudes. In particular, the orientor "Living choice", related to the issue of sustainability, has a U of 0.85, indicating that most farmers plan to continue with the land use/cover changes introduced by the GfG. The GfG has also improved farmers' environmental consciousness ("Coexistence", with a U of 0.99), which is likely to have a positive impact on its sustainability. Overall, Shi and Wang (2011) concluded that the GfG project had positive impacts in Mizhi County and the land uses/land cover changes brought about by the GfG will continue after the subsidies end.

In order to examine sustainability and forecast the farm household's post-contract land-use decisions, Uchida et al. (2005) directly asked households in Ningxia and Guizhou Province what they intended to do after program payments stopped (Fig. 13.1).[5] The central government required that 80 % of the land be planted with ecological trees and 20 % with economic trees. While the actual implementation in Guizhou Province was consistent with the government's requirement, the survey shows that more than 50 % of households stated that they would have preferred to plant economic trees. Uchida et al. (2005) argued that, because of the high proportion of ecological trees with limited economic benefits, there could be a greater danger of reconversion in the future when program payments cease. Thirty four percent of the participating farmers in Guizhou Province said that, if the government were to stop the payments after 5 years, they would shift their land back to cropping. Similarly, 29 % of the sample farmers in Ningxia Province stated that they had the same intentions (Uchida et al. 2005). On the other hand, Zhang et al. (2008b) found that 26 % of farmers in Ningxia planned to reconvert their land "for sure" and another 20.9 % "probably". The pressure to reconvert the land may be more serious

[5] Uchida et al. (2005) is based on a sample of 144 participating households from 16 randomly selected villages in Ningxia and Guizhou Provinces.

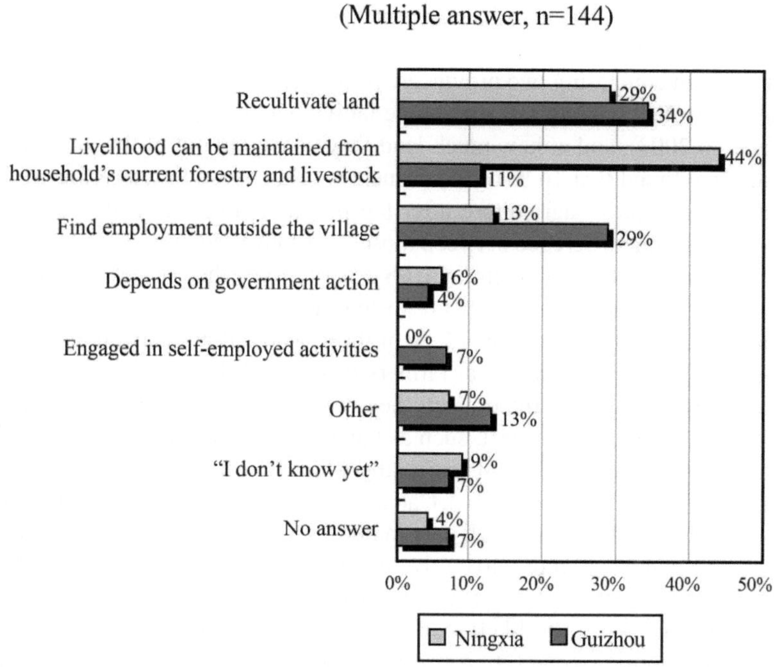


**Fig. 13.1** Summary of opinions of farm households about reconversion plans if GFG program payments stopped in Ningxia and Guizhou after 5 years, 2000 (Source: Uchida et al. 2005)

in Guizhou Province because the average land holdings per household is lower, and farmers may need to reconvert land back to agricultural production if they cannot find (or retain) alternative sources of income off-farm. In Ningxia Province 44 % of the farmers said that they believed their new mix of forestry and livestock enterprises would sustain their livelihood after the Grain for Green program. In contrast, only 11 % of the farmers in Guizhou Province replied that they would be able to do so. Not surprisingly, more farmers in Guizhou Province (29 %) replied that if payment were to stop, they would also seek off-farm jobs outside the village (versus 13 % in Ningxia Province). Hence, if the program encourages or pressures farmers to shift into activities that can provide them with incomes even after the program subsidies end, there is likely to be fewer pressures to return the set-aside land back to cultivation (Uchida et al. 2005).

Uchida et al. (2005) argued that the differences between the two provinces regarding the need for alternative off-farm jobs may also reflect the different economic environments that exist in the two provinces. First, as mentioned above, the average holdings of land per household in the sample are lower in Guizhou than in Ningxia. Although in both provinces more than 50 % of the sown area of households was set aside under the GfG, the amount of land remaining under cultivation is less, on average, for farmers in Guizhou Province. Therefore, those farmers have a greater need to find alternative sources of income outside the land-intensive agricultural sector. If the opportunities for off-farm employment dwindle after the program ends, it is plausible that farmers would revert the land to pre-GfG uses (Uchida et al. 2005).

Rather than comparing two provinces, Wang and MacLaren (2011) looked at the availability of land among farmers in one individual county (Dunhua County, Jilin Province, in 2003), and at how much land they converted out of the total land they owned. They found that 16 % of the farmers would choose to return the afforested and reforested land to agriculture after the program subsidies end. However, the survey shows a big difference between those who converted only some of their land, and those who converted all their land; 88.2 % of those who converted all their land claimed they wanted to reconvert afforested land to pre-GfG land uses after the program ends, compared to only 7.2 % of those who did not convert all their land. Although the surveys showed that farmers recognized the importance of the GfG and supported the aims of the project, they did not necessarily accept the personal costs associated with the project, such as the adverse impact that losing all of their croplands had on their livelihood, especially when their main source of income was farming (Wang and Maclaren 2011). Cao et al. (2009b) also found that while 63.8 % of the households in his fieldwork area in northern Shaanxi Province supported the GfG, 37.2 % planned to return to cultivating the converted forested areas and grassland, once the project's subsidies end in 2018.

One way to discern the likelihood of returning retired cropland to cultivation when the GfG subsidies end is to compare the wage rates for agricultural production and off-farm employment. Yin and Liu (2007) argued that these rates can be derived by dividing the net revenues from agricultural and off-farm employment by the corresponding labor times. That study revealed that the wage rates of participating households from off-farm opportunities were universally higher than those from agriculture for the years 2006–2008 (Fig. 13.2). It therefore can be inferred that rural laborers will prefer off-farm work. On the other hand, Fig. 13.2 also shows that the difference is decreasing. Yin and Liu (2007) argued that if this trend continues, it is likely that more rural laborers will revert to farming. Uchida et al. (2005) offer a counter-argument, contending that people will not return to farm work even if the

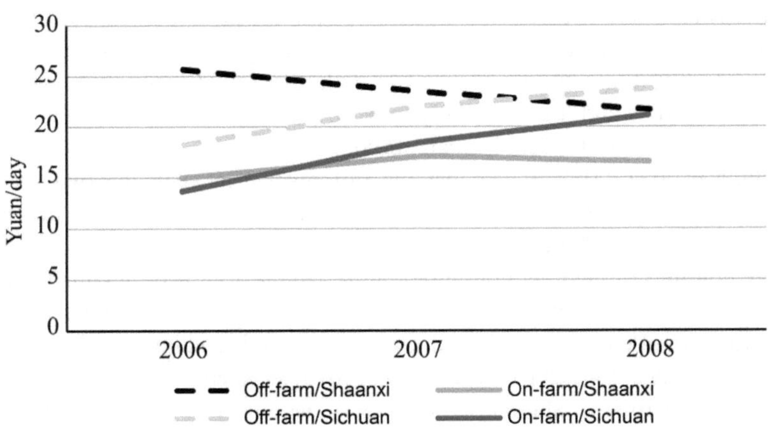


**Fig. 13.2** Estimated wage rates of different jobs for participants (Source: Yin and Liu 2007)

differences between on-farm and off-farm incomes decline, because when the farmers shift their land away from crop cultivation to other productive uses, they gradually increase the opportunity cost of reconversion.

## Property Rights of Land and Trees

Uchida et al. (2005: 78) point out that the CCICED (2002) raised concern that "uncertainty over the lack of property rights and the future responsibility for management of the trees may mean that farmers do not have strong incentives to maintain their forest plots in the long term. [...] Incentives to preserve natural resources and to invest in trees and other land improvements for future benefits will be hindered without well-established property rights, because the future benefits may not accrue to those who manage them. The uncertainty over the property rights of the trees planted under the program also may discourage the participating farmers from managing the trees, thereby diminishing the long-term environmental benefit of the program."

Grosjean and Kontoleon (2009) reached similar conclusions in their two-province study of farmers' choices when the GfG ends. Surveys were carried out in Ningxia Province, situated in northwest China into the middle reaches of the Yellow River, and Guizhou Province, located in the southwest on the reaches of the Yangtze River. These provinces were selected because they were among the first where the GfG was implemented, and because their particularly poor economic and ecological conditions relative to the rest of China were envisaged to provide particularly important information for the sustainability of the GfG (Grosjean and Kontoleon 2009).[6]

In order to assess the viability of the program in its current form, Grosjean and Kontoleon (2009) analyzed responses to contingent behavior questions over household land and labor allocation intentions after the program ends, under three plausible and mutually exclusive alternative post-GfG scenarios: (1) the program will be renewed in its current form; (2) the program will be terminated; or (3) a different and new program will be introduced. The first two choices were naturally confined to GfG participants alone and were focused on both labor and land allocation intentions of participating households. For the third scenario they used a choice experiment in which both participants and non-participants were asked to select their preferred policy option from a range of hypothetical land set-aside policies (Grosjean and Kontoleon 2009).

For the first scenario, where the program is renewed, 63 % of farmers said they would sign up for the program and maintain or increase reforested land, while 42 % said they would decrease their on-farm labor activities. For the second scenario, where the program is terminated, only 38 % of farmers said they would continue to maintain

[6] Both household and village level data were collected via in-person interviews with the head or spouse of randomly selected households (without replacement) and with village leaders. Household data were collected for both GfG participants and non-participants. In total, 286 households in 44 villages were surveyed (Grosjean and Kontoleon 2009).

their reforested lands, while 67 % said that they would increase their on-farm labor activities (Grosjean and Kontoleon 2009). These results, in addition to the analysis derived from the third scenario, were used to reach the following conclusions:

First, the GfG should address the root causes of households' inefficient allocation of resources, in particular uncertain property rights and high costs of labor mobility, in dealing with the underlying problem of an oversupply of farm labor. Second, in cases where the GfG is renewed, Grosjean and Kontoleon (2009) recommend providing better forestry training to participating households along with more autonomy in managing their trees. Third, in the event that subsidies are not renewed, farmers will tend not to reconvert their reforested lands, provided that the commercial value of the reforested trees is high. Fourth, secure property rights were also shown to be important in the post-GfG scenario when subsidies are terminated. Since subsidies will end sooner or later, offering farmers secure property rights seems to be an important issue. Fourth, in the scenario where a new program is offered, Grosjean and Kontoleon (2009) found that the likelihood of enrollment would be affected, not just by the level of subsidies, by the accessibility and attraction of off-farm employment (e.g. creating employment centers, reducing local travel costs, enhancing education, and the level of the off-farm wages), and by wider institutional reforms that would include land tenure, land renting, and land management.

Making a somewhat similar argument to that of Grosjean and Kontoleon (2009), Groom et al. (2010, in Grosjean and Kontoleon 2009) found that, almost 10 years after the GfG started, market and institutional constraints (primarily incomplete property rights and high transactions costs) still constituted serious impediments to the reallocation of labor toward off-farm activities, and thus remained important contributors to the vicious cycle of inefficient production processes, poverty, and environmental degradation.

Land tenure and exchange rights have been shown to be essential determinants of agricultural and labor allocation choices in China. In particular, insecure land rights may discourage households from committing to land quality investments (such as the maintenance of reforested trees) while they may also constrain household members from seeking more profitable off-farm employment opportunities due to the fear of losing unused land. Therefore, land tenure and exchange rights can be expected to have a significant impact on the likelihood of converting land to pre-GfG uses (Grosjean and Kontoleon 2009).

**Conclusions**

This chapter has reviewed the literature on various issues related to the sustainability of the GfG. Sustainability in this case relates to the question of whether households will maintain the land cover/land use changes introduced by the GfG, or whether they will revert to pre-GfG land uses once the subsidies end. When the program began, the economic returns of land

(continued)

and labor from reforestation were substantially lower than those generated by grain or cash crop production. That difference was covered by subsidies, and the ability to engage in work on other land, or off-farm, made the program attractive to many farmers. However, payments will eventually cease, and over the longer term farmers will revert their land back to pre-GfG uses if the incomes from trees are not competitive, or if there are insufficient off-farm opportunities.

Few studies tried to estimate the post-subsidies economic benefits from GfG land-use changes, but those that did concluded that in many cases (at least for economic trees) revenues from the new land uses were superior to those of pre-GfG land use. However, this chapter, like previous ones, has shown that researchers' findings display considerable variation. This is not surprising. It is unreasonable to expect that the same results will be found across all of China. China is a very diverse country, economically, socially and ecologically, and a program that was rather homogeneous throughout the country (and indeed has been criticised as such when, for example, promoting tree planting in arid areas) cannot be expected to fare similarly everywhere. Hence, some researchers found that a majority of households plan to maintain the new land use/land cover, while others have found that they will not. Most research presented is localised in a relatively small area, and the differences described may simply be due to local variations in environmental and socio-economic conditions.

While most studies focused on comparing the economic benefits of pre-GfG land uses to those of post-GfG land uses, it is also clear that the situation in many places have changed, so that comparison may be irrelevant. In particular, in many cases, as the previous chapters have also shown, the most productive members of many families have migrated out of the rural areas, and only aged people and children remain in the villages. For them, it might be difficult to grow food crops while it is possible to harvest economic trees. Thus, income is not the only consideration that may affect land use.

As the program is set to end starting in 2015, it is disconcerting that a nation-wide survey of the attitudes of farmers towards the GfG, as well as a comparison of the economic returns of economic and ecological trees to pre-GfG land uses, has not been undertaken recently. In our opinion, there is insufficient evidence to determine how sustainable the GfG will be, once the subsidies end. It is quite possible that unless a transition to post-GfG subsidies is properly planned, many of the positive impacts of the GfG will be lost, and a sizable proportion of the investment made (some Yuan 430 billion) will have been squandered

# References


ACCA21. (1994). *China's Agenda 21-white paper on China's population, environment and development in the 21st century*. Beijing: The Administrative Centre for China's Agenda 21.

Alig, R., Adams, D., McCarl, B., Calloway, J. M., & Winnett, S. (1997). Assessing effects of mitigation strategies for global climate change with an inter-temporal model of the US Forest and Agricultural Sectors. *Environmental and Resource Economics, 9*(3), 259–274.

Anonymous. (1998). *Mizhi Prefecture statistical year book*. Mizhi Prefecture.

Anonymous. (2002). *Jianyang County statistical yearbook*. Jianyang Prefecture.

Babcock, B. A., Lakshminarayan, P. G., Wu, J. J., & Zilberman, D. (1996). The economics of a public fund for environmental amenities: A study of CRP contracts. *American Journal of Agricultural Economics, 78*(4), 961–971.

Bardhan, P., & Udry, C. (1999). *Development economics*. Oxford: Oxford University Press.

Bennett, M. T. (2008). China's Sloping Land Conversion Program: Institutional innovation or business as usual? *Ecological Economics, 65*(4), 699–711.

Bennett, M. T., & Xu, J. (2005, June 15–18). *China's Sloping Land Conversion Program: Institutional innovation or business as usual?* Paper presented at the workshop on "Payments for Environmental Services (PES) – Methods and design in developing and developed countries", Titisee, Germany.

Bennett, M. T., Zhang, L., Dai, G., Xie, C., Zhao, J., et al. (2004). *A review of the Program for Conversion of Cropland to Forest and Grassland* (Sustainable land use change in the North West Provinces of China Research Reports). Canberra: Australian Centre for International Agricultural Research.

Bennett, M. T., Metha, A., & Xu, J. (2011). Incomplete property rights, exposure to markets and the provision of environmental services in China. *China Economic Review, 22*, 485–498.

Bossel, H. (1999). *Indicators for sustainable development: Theory, method, applications*. Winnipeg: IISD.

Brandt, L., Huang, J. K., Li, G., & Rozelle, S. (2002). Land rights in rural China: Facts, fictions and issues. *The China Journal, 47*(1), 67–97.

Brandt, L., et al. (2008). China's great transformation. In L. Brandt & G. T. Rawski (Eds.), *China's great economic transformation*. Cambridge: Cambridge University Press.

Cai, Y. L., Fu, Z. Q., & Dai, E. F. (2002). The minimum area per capita of cultivated land and its implication for the optimization of land resource allocation. *Acta Geographica Sinica, 57*(2), 127–134.

Cao, S. (2008). Why large-scale afforestation efforts in China have failed to solve the desertification problem. *Environmental Science & Technology, 42*(6), 1826–1831.





C.O. Delang, Z. Yuan, *China's Grain for Green Program*,
DOI 10.1007/978-3-319-11505-4

Cao, S., Chen, J., Chen, L., & Gao, W. (2007a). 退耕还林项目对陕北地区自然与社会的影响 [Impact of Grain for Green Project to Nature and Society in North Shaanxi of China]. *Scientia Agricultura Sinica, 40*, 972–979.

Cao, S., Chen, L., Xu, C., & Liu, Z. (2007b). Impact of three soil types on afforestation in China's Loess Plateau: Growth and survival of six tree species and their effects on soil properties. *Landscape and Urban Planning, 83*(2), 208–217.

Cao, S., Chen, L., & Yu, X. (2009a). Impact of China's Grain for Green Project on the landscape of vulnerable arid and semi-arid agricultural regions: A case study in northern Shaanxi Province. *Journal of Applied Ecology, 46*(3), 536–543.

Cao, S., Xu, C., Chen, L., & Wang, X. (2009b). Attitudes of farmers in China's northern Shaanxi Province towards the land-use changes required under the Grain for Green Project, and implications for the project's success. *Land Use Policy, 26*(4), 1182–1194.

CCICED. (2002). *Implementing the Natural Forest Protection Program and the Sloping Land Conversion Program: Experiences and Lessons*, Preliminary Draft.72.

CFY. (1987). *China Forestry Yearbook (1949–1986)*. Beijing: Ministry of Forestry, China Forestry Publishing House.

CFY. (1999). *State Forestry Administration. China Forestry Yearbook 1998*. Beijing: The Publishing House of Forestry (in Chinese).

Chen, J. (1999). *Chinese law: Towards an understanding of Chinese law: Its nature and development*. Hague: Kluwer Law International.

Chen, B. M. (2001). *The integrated agricultural resources production capability and population supporting capacity in China*. Beijing: Meteorological Press.

Chen, N. B. (2010). *State, market and life chances: Evidence from rural Guangdong China (1978–2004)*. Beijing: Central Compilation & Translation Press (in Chinese).

Chen, Z. H., & Shou, X. L. (2001). China s environmental diplomacy: Historical transitions and current challenges. *China Population, Resources and Environment, 11*(4), 52–56 (in Chinese).

Chen, X., Zhang, X., Zhang, Y., & Wan, C. (2009). Carbon sequestration potential of the stands under the Grain for Green Program in Yunnan Province, China. *Forest Ecology and Management, 258*, 199–206.

China Daily. (2000). *New laws to ensure better environment*. http://www.china.org.cn/e-15/15-3-b/15-3-b-5.htm. Accessed 10 Jan 2014.

China Daily. (2004). *Reforestation to continue in China's West*. http://www.china.org.cn/archive/2004-10/15/content_1109508.htm. Accessed 10 Sept 2013.

CLC. (1982a) (1949–1950). *Central Legislation Committee. Chinese people's government regulations assembly (1949–1950)*. Beijing: The Law Publishing House.

CLC. (1982b) (1957). *Central Legislation Committee. Chinese People's Government Regulations Assembly (1957)*. Beijing: The Law Publishing House.

Cooper, J. C., & Osborn, T. (1998). The effect of rental rates on the extension of Conservation Reserve Program contracts. *American Journal of Agricultural Economics, 80*(1), 184–194.

CSY. (1991). *China statistical yearbook 1991*. Beijing: China Statistical Publishing House.

Cui, H. (2009). 退耕还林工程社会影响评价理论及实证研究 [Theoretical and practical research on social impacts of the Grain for Green Program]. Beijing: Intellectual Property Publishing House.

deBrauw, A. (2002). *Three essays on migration, education and household development in rural China* (p. 67). Davis: University of California.

Delang, C. O., & Wang, W. (2012). Chinese forest policies in the age of decentralization (1978–1997). *International Forestry Review, 14*(1), 13–26.

Delang, C. O., & Wang, W. (2013). Chinese forest policy reforms after 1998: The case of the Natural Forest Protection Program and Slope Land Conversion Program. *International Forestry Review, 15*(3), 290–304.

Démurger, S., Fournier, M., & Shen, G. Z. (2005). Forest protection policies: National guidelines and their local implementation in northern Sichuan. *China Perspectives, 59*, 2–15.

Démurger, S., Hou, Y. Z., & Yang, W. Y. (2009). Forest management policies and resource balance in China: An assessment of the current situation. *The Journal of Environment and Development, 18*(1), 17–41.

Dingbian GFG Office. (2007). *A summary report of implementation of the Sloping Land Conversion Program*. Dingbian County: Shaanxi Province.

Dingbian Statistics Bureau. (2006). *Statistics Yearbook of Dingbian County 2005*. Dingbian County: Shaanxi Province

DOF (Department of Forestry, China). (1995). *The forestry executive plan in 21st century agenda in Chi.*

Dong, X. (1996). Two-tier land system and sustained economic growth in post-1978 rural China. *World Development, 24*(5), 915–928.

Dong, J. (2003). 坡耕地与林地价值比较研究——兼论退耕还林的效益 [Value comparison between sloping fields and forestland – Discussing the benefits of converting from farmland to forest simultaneously]. *China Population, Resources and Environment, 13*, 86–88.

Dong, J., Liu, J., & Shi, W. (2010). China's Slope Land Conversion Program at the beginning of 21st century and its habitat suitability in typical region of Loess Plateau. *Journal of Resources and Ecology, 1*(1), 36–44.

Du, S. F. (2001). *Environmental economics*. Beijing: Encyclopedia Press.

Du, R. (2002). *Contemporary China's agricultural cooperation*. Beijing: Contemporary China Publishing House.

Du, Y. (2012). Duty crime report in agriculture, forestry, water department. *Journal of Guangxi University (Philosophy and Social Science), 34*(4), 39–50 (in Chinese).

Earth Policy Institute. (2013a). *Grain and soybean area harvested in China, 1960–2011*. http://www.earth-policy.org/data_center/C24. Accessed 10 Jan 2014.

Earth Policy Institute. (2013b). *Grain production, consumption, imports and exports in China, 1960–2010*. http://www.earth-policy.org/data_center/C24. Accessed 10 Jan 2014.

Fan, C. C. (2006). China's Eleventh Five-Year Plan (2006–2010): From "Getting Rich First" to "Common Prosperity". *Eurasian Geography and Economics, 47*(6), 708–723.

FDOGX (Forestry Department of Guang Xi). 2006. *The comparison on conversion of cropland to forest and grassland: From grain to cash*. Accessed 5 Jan 2013 at http://www.msgnet.gov.cn/news-263.aspx (in Chinese).

Feder, G., et al. (1990). The relationship between credit and productivity in Chinese agriculture: A microeconomic model of disequilibrium. *American Journal of Agricultural Economics, 72*(4), 1151–1157.

Feng, C. (2005). Discussion on private forestry development in China. *Chinese Forestry Science and Technology, 4*(3), 90–93. http://www.sfmchina.cn/issuance/UploadFile/2007112802.pdf. Accessed 10 Jan 2014.

Feng, Z. M., & Zhang, P. T. (2002) Xibu diqu tuigeng de gongzhong diaocha fenxi [Public inquisition of converting cultivated land to forests and grass in west China]. In *Di san jie haixia liang an sandi shuitu ziyuan shengtaihuanjing baoyu yantaohui* [Proceedings of the 3rd symposium on water, land resources and ecological environment of the Mainland, Hong Kong and Taiwan of China], Taipei, pp. 251–255.

Feng, Z., Liu, B., & Yang, Y. (2005a). A study of the changing trend of Chinese cultivated land amount and data reconstructing: 1949–2003. *Natural Resources Journal, 20*(1), 35–44 (in Chinese).

Feng, Z., Yang, Y., Zhang, Y., Zhang, P., & Li, Y. (2005b). Grain-for-green policy and its impacts on grain supply in West China. *Land Use Policy, 22*, 301–312.

Forest and Grassland Taskforce of China. (2003, January). In pursuit of a sustainable green west. *Newsletter*, Beijing.

FRSOC. (1983). *Forest Resources Statistics of China (1977–1981)*. Beijing: Ministry of Forestry (in Chinese).

FRSOC. (1989). *Forest Resources Statistics of China (1984–1988)*. Beijing: Ministry of Forestry (in Chinese).

Fu, G., Chen, S., Liu, C., & Shepard, D. (2004). Hydro-climatic trends of the yellow river basin for the last 50 years. *Climatic Change, 65*(1–2), 149–178.

Fukao, Y. (2000). Kodokogen no mura: Mizu, tsuchi, hito no ryushutu/saisei heno kokoromi (Japanese). In R. Kojima et al. (Eds.), *The structure of contemporary China* (pp. 267–310). Tokyo: University of Tokyo Press.

Gauvin, C., Uchida, E., Rozelle, S., Xu, J., & Zhan, J. (2010). Cost-effectiveness of payments for ecosystem services with dual goals of environment and poverty alleviation. *Environmental Management, 45*(3), 488–501.

Ge, W., Li, L., & Li, Y. (2006). 退耕还林工程可持续性问题探讨——对陕西省吴旗县、志丹县实施退耕还林的调查与思考 [Analysis on the sustainability about Grain to Green Program – A field survey of Grain to Green Program in Wuqi and Zhidan Counties of Shaanxi]. *Forestry Economics, 11*, 33–49.

Gong, Y., & Xu, J. (2004). Farmland conversion in Dafang County, Guizhou Province. In J. Xu & U. Schmitt (Eds.), *Case studies – CCICED Task Force on Forests and Grasslands, 340*. Beijing: China Forestry Publishing House.

Groom, B., Grosjean, P., Kontoleon, A., Swanson, T., & Zhang, S. (2010). Relaxing rural constraints: A "win-win" policy for poverty and environment in China. *Oxford Economic Papers, 62*, 132–156.

Grosjean, P., & Kontoleon, A. (2009). How sustainable are sustainable development programs? The case of the Sloping Land Conversion Program in China. *World Development, 37*(1), 268–285.

Guo, X., Gan, T., Li, S., & Luo, H. (2005). 退耕还林工程:问题、原因与政策建议 — 四川省天全县100戶退耕还林农戶的跟踪调查 [Assessment of socio-economic benefits of the Cropland Conversion to Forestland Program in Xinping County]. *China Rural Survey, 3*, 72–79.

He, D., & Zhu, D. (2010). Studies on collective forest tenure reform in the past 30 years. *Forestry Economics, 5*, 13–24.

Hoff, K., & Stiglitz, J. E. (1990). Imperfect information and rural credit markets – Puzzles and policy perspectives. *World Bank Economic Review, 4*, 235–250.

Hong, Y., & Li, X. (2000). Cultivated land and food supply in China. *Land Use Policy, 17*(2), 73–88.

Hori, S., & Kojima, K. (2008). The impact of the Sloping Land Conversion Program on rural area in China: A case study in Yulin district. *Japan Society of Tropical Ecology Tropics, 17*(2), 169–184.

Hsu, C. (2002). *China's modern history* (second volume). Hong Kong: The Chinese University of Hong Kong Press.

Hu, A. (2007). *China political and economic history (1949–1976)*. Beijing: Tsinghua University Press.

Huachi GfG Office. (2007). *A summary report of implementation of the Sloping Land Conversion Program in Huachi County*. Huachi: GfG Office.

Huachi Statistics Bureau. (2006). *Statistics yearbook of Huachi County 2005*. Huachi County: Gansu Province

Huang, J. (2000). *Land degradation in China: Erosion and salinity component* (CCAP Working Paper WP-00-E17). Beijing: Center for Chinese Agricultural Policy.

Huang, R. (2006). *Revolution and country: Research on the Rural Land Property 1949–1983: Case study in Xinzhou county in Hubei province*. Shanghai: Shanghai Academy of Social Science Press.

Huang, D., Yu, Z., & Wang, X. (1992). *The agricultural cooperation historical documents assembly since the establishment of China*. Beijing: The History Press

Huang, D., Li, B. Y., Yu, B. C., & Jiang, X. L. (2009). A report for monitoring and assessment of the socio-economic impacts of conversion cropland to forest project. *Forestry Economics, 9*, 65–73.

Jiang, Y., Lin, Y. S., & Sun, X. P. (2013). Green development and ecological construction. In *China green development index report 2011* (pp. 237–261). Heidelberg: Springer.

Jiao, J. (2005). 西北地区退耕还林工程的实施和政策落实 [Implementation and policy execution of removal lands from cultivation for afforestation project in the northwestern areas]. *Science of Soil and Water Conservation, 3*, 4–9.

International Fund for Agricultural Development. (2001). *Rural financial services in China: Thematic study*. Rome: International Fund for Agricultural Development.

Ke, S. (2007). *Empirical study and behaviour theory of household participation in the Grain for Green Program*. Beijing: China Agriculture Press.

Knight, J., & Song, L. (2005). *Towards a labour market in China*. Oxford: Oxford University Press.

Lai, H. Y. (2002). China's Western Development Program: Its rationale, implementation, and prospects. *Modern China, 28*(4), 432–466.

Lai, D. S., Cai, N., & Rong, T. T. (2013). Government green investment. In *China green development index report 2011* (pp. 317–350). Heidelberg: Springer.

Lei, J. (2002). *China's implementation of Six Key Forestry Programs. N.P.* http://www.china.org.cn/e-news/news02-05-14.htm. Accessed 10 Nov 2013.

Lei, J., & Zhu, L. (2002). *China's implementation of Six Key Forestry Programs*. China Tibet Information Center/State Forestry Administration. Beijing.

Lenin, V. (1972). *Selected works of Lenin* (Vol. 4). Beijing: People Publishing House.

Li, T. (1985). *Modern China's forestry*. Beijing: China Social Science Press.

Li, F. (1988a). *My memories of forestry construction*. Beijing: China Forestry Publishing House.

Li, J. (1988b). *Chinese resources and environment*. Beijing: Xinhua Publishing House.

Li, Z. (2001). *Conserving natural forests in China: Historical perspective and strategic measures* (Working Report). Beijing: Chinese Academy of Social Sciences.

Li, S. D. (2002). Comparison on conversion of cropland to forest and grassland in the world. *World Forestry Research, 15*(2), 22–27.

Li, S. D. (2003). Study on the development stages of converting cropland for forest and grassland. *World Forestry Research, 16*(1), 36–41 (in Chinese).

Li, S. (2004a). 中国退耕还林研究 [Research on conversion of farmland to forests in China]. Beijing: Science Press.

Li, W. (2004b). Degradation and restoration of forest ecosystems in China. *Forest Ecology and Management, 201*(1), 33–41.

Li, Y. C. (2005). *The Grain for Green Project in China*. Beijing: China Forestry Press.

Li, Y. C. (2007). *Heightened concerns to the slope land conversion program from central media*. Accessed 5 Jan 2013 at http://www.hnly.gov.cn/portal/lydt/gngj/webinfo/2007/10/1257325230072381.htm (in Chinese).

Li, C. (2009a). *China's agricultural tax history*. Beijing: State Administration of Taxation. http://www.ctax.org.cn/news/rdzt/ctax60/item/item2_7.html. Accessed 5 Oct 2013 (in Chinese).

Li, X. (2009b). 退耕还林理论基础及林草模式的实践应用 [Basic theory and practice of reforestation of Grain for Green Program]. Beijing: Science Press.

Li, Y. (2009c, December 26–28). Social and ecological impact of PES program in arid region: The case from Zhang-ye in Northwest China. In *The 1st international conference on information science and engineering*. Nanjing University of Information Science & Technology.

Li, Z. B., & Lu, Y. (2004). *Adjust the comparison on conversion of cropland to forest and grassland: From grain to cash*. http://news.xinhuanet.com/st/2004-09/24/content_2014735.htm (in Chinese).

Li, S., & Zhai, H. (2002). 世界林业生态工程对比研究 [The comparison study on forestry ecological projects in the world]. *Acta Ecologica Sinica, 22*, 1976–1982.

Li, D. Y., Bo, F. M., & Tao, J. L. (2006). 湖南省退耕还林工程成效与发展对策 [The achievements, effects and development countermeasures of the Returning Land for Farming to Forestry Project in Hunan]. *Hunan Forestry Science & Technology, 33*, 1–5.

Li, M., Liu, A., Zou, C., Xu, W., Shimizu, H., & Wang, K. Y. (2012). An overview of the "Three-North" Shelterbelt project in China. *Forestry Studies in China, 14*(1), 70–79.

Liang, W., Bai, C., Sun, B., Hao, D., & Qi, J. (2006). 黄土丘陵沟壑区退耕还林(草)区土壤水分-物理性质研究 [Study on Soil Moisture-Physical Properties of Returning Land for Farming to Forestry(Grass) Region in Gullied Rolling Loess Area]. *Soil and Water Conservation in China, 3*, 17–18.

Liang, Y., Li, S., Feldman, M. W., & Daily, G. C. (2012). Does household composition matter? The impact of the Grain for Green Program on rural livelihoods in China. *Ecological Economics, 75*(Issue C), 152–160.

Liao, X. C., & Zhang, Y. Q. (2008). Economic impacts of shifting sloping farm lands to alternative uses. *Agricultural Systems, 97*(1–2), 48–55.

Liu, J. (2006). *Forests in the mist*. Ph.D. thesis, Wageningen University, Wageningen.

Liu, C. (2007). The collective forest property right system analysis, change and performance – Arrangement (3). *Forestry Economy*, (2), 45–51 (in Chinese).

Liu, S. (2010). Studies on returning and paying compensations in the rural areas of Jiangsu province in the period of national economic readjustment. *CPC Research, 3*, 13–18 (in Chinese).

Liu, J., & Diamond, J. (2005, June 30) China's environment in a globalizing world. *Nature, 435*, 1179–1186.

Liu, C., & Li, H. (2010). 中国退耕还林政策系统性评估研究 [Study of the Systematic Assessment of Conversion Program of cropland to forest in China]. Beijing: Economy & Management Publishing House.

Liu, J., & Yuan, J. (2007). China's boom in household management of forests. *FAO: Unasylva, 228*(58), 19–22. http://www.fao.org/docrep/010/a1346e/a1346e05.htm. Accessed 10 Jan 2014.

Liu, Y., & Zhou, Q. (2005). 退耕还林政策的激励机制缺陷 [The institutional defects of de-farming and reforestation policy]. *China Population, Resources and Environment, 15*, 108–111.

Liu, S., Carter, M., & Yang, Y. (1998). Dimensions and diversity of property rights in rural China: Dilemmas on the road to further reform. *World Development, 26*(10), 1789–1806.

Liu, F., Huang, C., He, T., Qian, X., Liu, Y., et al. (2002). 黄壤旱坡地退耕还林还草对减少土壤磷流失的作用 (Roles of reducing phosphorus loss of surface runoff from yellow soil in hilly areas by de-farming and reafforestation). *Journal of Soil Water Conservation, 3*, 20–23.

Liu, J., Li, S., Ouyang, Z., Tam, C., & Chen, X. (2008). Ecological and socioeconomic effects of China's policies for ecosystem services. *Proceedings of the National Academy of Sciences, 105*(28), 9477–9482.

Liu, J. Y., Zhang, Z. X., Xu, X. L., & Kuang, W. (2009). Analysis of spatial pattern of land use change and its driving force in China in the early of 21st century. *Acta Geographica Sinica, 64*(12), 1411–1420 (in Chinese).

Liu, C., Lu, J. Z., & Yin, R. S. (2010). An estimation of the effects of China's Priority Forestry Programs on-farmers' income. *Environmental Management, 45*(3), 526–540.

Lok, W. K. (2009). The relationship between central and local governments under the unitary state system of China. In J. C. Oliveira & P. Cardinal (Eds.), *One country, two systems, three legal orders – Perspectives of evolution* (pp. 527–540). Heidelberg: Springer.

Long, H., & Zou, J. (2010). Grain production driven by variations in farmland use in China: An analysis of security patterns. *Journal of Resources and Ecology, 1*(1), 60–67.

Long, H., Liu, Y., Li, X., & Chen, Y. (2010). Building new country side in China: A geographical perspective. *Land Use Policy, 27*(2), 457–470.

Lu, Y. H. (2011). International convention related to China's forestry. *Forestry of Guangxi, 2011*(4), 53 (in Chinese).

Lu, W., Landell-Mills, N., Liu, J., Xu, J., & Liu, C. (2002). *Getting the private sector to work for the public good – Instruments for sustainable private sector forestry in China. Instruments for sustainable private sector forestry series*. London: International Institute for Environment and Development.

Luo, H. B., Qian, X. G., & Liu, F. (2003). 喀斯特山區退耕還林(草)保持水土生態效益研究 [Ecological benefit of soil and water conservation in hilly areas by de-farming and re-afforestation]. *Journal of Soil and Water Conservation, 17*(4).

Luo, L. H., Hu T. X., & Wan, X. Q. (2006). 天全县几种退耕还林类型林地土壤理化性质年际动态变化研究 [Dynamic variation of physical and chemical properties of soil in different patterns of returning to forest]. *Journal of Zhejiang Forestry Science and Technology, 26*(1), 18–22.

Ma, E. (1991). From three-fixed to forestry joint – The dual operation system. *China's Rural Economy, 7*, 45–47.

Ma, Y., & Fan, S. (2005). 沙漠化地区退耕还林政策的生态经济效应分析——以民勤县为例 [Eco-economic effect of actualizing de-farming and reafforestation policy in desertification areas: Taking Minqin County as a Case]. *Journal of Natural Resources, 20*, 590–596.

Macfarquhar, R., & Fairbank, J. K. (1978a). *The people's republic part 1: The emergence of revolutionary China 1949–1965* (The Cambridge History of China, Vol. 14). Cambridge, UK: Cambridge University Press.

Macfarquhar, R., & Fairbank, J. K. (1978b). *The people's republic part 2: Revolutions within the Chinese revolution 1966–1982* (The Cambridge History of China, Vol. 15). Cambridge, UK: Cambridge University Press.
Mao, Y. S., Zhao, N., & Yang, X. J. (2013). *Food security and farm land protection in China.* Singapore: World Scientific Pub. p. 30.
Mcmillan, J., Whalley, J., & Zhu, L. (1989). The impact of China's Economic reforms on agricultural productivity. *The Journal of Political Economy, 97*(4), 781–807.
Mei, Z. M., & Xiong, K. N. (2003). A study on the dynamic characteristics of soil erosion and eco-benefit evaluation in Karst region-A case study on Qingzhen demonstrating site of returning farmland to wood(grass), Guizhou. *Carsologica Sinica, 22*, 136–143.
Meng, M., Feng, Y., & Xun, J. (2000). *Population and sustainable development in Yunnan.* Consultancy report written for the Yunnan provincial government (387pp.) (in Chinese).
Menzies, N. (1988). Three hundred years of Taungya: A sustainable system of forestry in Southern China. *Human Ecology, 16*(4), 361–376.
Ministry of Land and Resources. (2004). *2003 China National Report on Land and Resources.* Beijing: Geological Publishing House.
MOYN. (2009). *The compilation of conversion of cropland to forest and grassland project of Yunnan, Kunming, China.* Unpublished report by the Yunnan Forestry Department, Kunming, China for the Management Office for the Conversion of Cropland to Forest and Grassland Project of Yunnan Forestry Department (258pp.) (in Chinese).
National Bureau of Statistics. (2009). *China statistical yearbook.* Beijing: National Bureau of Statistics.
National Bureau of Statistics of China. (1990–2004). *China statistical yearbook.* Beijing: China Statistics Press.
National Bureau of Statistics of China. (1996–2012). *Chinese statistical yearbook.* Beijing: China Statistical Press.
National Development and Reform Commission. (2008). Report on the implementation of the 2007 Plan for National Economic and Social Development and on the 2008 Draft Plan for National Economic and Social Development.
NBS. (2004). *National bureau of statistics of China rural economic and social investigation team* (China Rural Poverty Monitoring Report 2004). Beijing: China Statistical Press (in Chinese).
Normile, D. (2007). Getting at the roots of killer dust storms. *Science, 317*(5836), 314–316.
Nyberg, A., & Rozelle, S. (1999). *Accelerating China's rural transformation.* Washington, DC: The World Bank.
O'Connor, J. (2000). *Natural causes: Essays in ecological Marxism* (Chinese translation by Z. D. Tang & P. H. Zang). Nanjing: Nanjing University Press (in Chinese).
Ostwald, M., Moberg, J., Persson, M., & Xu, J. (2011). *The Chinese Grain for Green Program – Assessing the sequestered carbon from the land reform.* Sweden: World Renewable Energy Congress.
Pan, L. T. (2006). *The 10th Five-Year Plan (2001–2005) Chinese Government official Web portal.* Downloaded 10 January 2014 from http://english.gov.cn/2006-04/05/content_245624.htm
Pan, Y., Lou, T., Birdsey, R., Hom, J., & Melillo, J. (2004). New estimation of carbon storage and sequestration in China's forest: Effects of age-class and method on inventory-based carbon estimation. *Climatic Change, 67*, 211–236.
Pan, L., Shi, Y. H., Xiong, Y. P., Wang, Z. Q., Xiang, Z. D., & Ma, D. J. (2006). 秭归县退耕还林水源涵养效益计量 [Quantitative evaluation on water conservation benefit of converting land for forest in Zigui County]. *Hubei Forestry Science and Technology, 3*, 1–4.
Peng, H. G. C., Xu, Z., Yin, Y., & Xu, W. (2007). Social, economic, and ecological impacts of the "Grain for Green" project in China: A preliminary case in Zhangye, Northwest China. *Journal of Environmental Management, 85*(3), 774–784.
Richardson, S. D. (1966). *Forestry in communist China.* Baltimore: Johns Hopkins Press.
Richardson, S. D. (1990). *Forests and forestry in China: Changing patterns of resource development.* Washington, DC: Island Press.

Rozelle, S., Huang, J., Husain, S. A., & Zazueta, A. (2000). *China from afforestation to poverty alleviation and natural forest management: Evaluation country case study series*. Washington, DC: The World Bank.

Rozelle, S., Brandt, L., Guo, L., & Huang, J. K. (2001). Land rights in China: Facts, fictions, and issues. *China Journal, 47*(1), 67–97.

SAC. (1981). *State Agricultural Commission agricultural cooperation important documents assembly (1949–1957)*. Beijing: The Central School of the Communist Press.

Shen, Z. (2008). *Reflections and choices: The consciousness of Chinese intellectuals and the anti-rightist campaign (1956–1957)* (The History of the People's Republic of China, Vol. 3). Hong Kong: Research Centre Contemporary Chinese Culture/The Chinese University of Hong Kong.

Shen, Y. C., Yang, Q. Y., Jing, K., & Xu, J. X. (2003). Thoughts and proposals on speeding up water and soil conservation and ecological environment construction on the Loess Plateau. *Science and Technology Review, 4*, 55–59.

Shi, W., & Wang, K. (2011). Assessment of ecological, economic and social impacts of grain for green on the counties of north Shaanxi in the Loess Plateau, China: A case study of Mizhi County. *African Journal of Biotechnology, 10*(70), 15763–15769.

Skaggs, R. K., Kirksey, R. E., & Harper, W. M. (1994). Determinants and implications of post-CRP land use decisions. *Journal of Agricultural and Resource Economics, 19*(2), 299–312.

Smil, V. (1993). Afforestation in China. In A. Mather (Ed.), *Afforestation: Policies, planning and progress*. London: Belhaven Press.

Stanturf, J. A., Madsen, P., & Lamb, D. (2012). *A goal-oriented approach to forest landscape restoration*. Heidelberg: Springer.

State Administration of Grain. (2002). *Grain production*. Accessed online 28 Apr 2014 at http://www.chinagrain.gov.cn/english/Grain%20Production.html

State Council of China. (2002a). *Regulations of restoring farmland to forest*. http://xzzf.forestry.gov.cn/portal/zfs/s/3273/content-489417.html. Accessed 10 Jan 2014 (in Chinese).

State Council of China. (2002b). 国务院关于进一步完善退耕还林政策措施的若干意见 [Improvement of the subsidy scheme for the grain for green program]. http://www.gov.cn/gongbao/content/2002/content_61463.htm

State Council of China. (2004). 国务院办公厅关于完善退耕还林粮食补助办法的通知 [Notice to improve the grain subsidy of the grain for green program]. http://tghl.forestry.gov.cn/portal/tghl/s/2166/content-448771.html

State Council of China. (2007). 国务院关于完善退耕还林政策的通知 [Notice of improving grain for green program policy]. Accessed online 28 Apr 2014 at http://www.gov.cn/zwgk/2007-08/14/content_716617.htm

State Forestry Administration. (1999). *State Forestry Administration. Forest development in China (1949–1999)*. Beijing: The Publishing House of Forestry (in Chinese).

State Forestry Administration. (2000a). 1958 年林业大事记 [The chronology of the forestry events in 1958]. http://www.forestry.gov.cn/main/60/content-114.html

State Forestry Administration. (2000b). *China forestry development annual report*. Beijing: China Forestry Press.

State Forestry Administration. (2000c). 第五次全国森林资源清查主要结果 (1994–1998年) [The fifth national forest inventory report 1994–1998]. http://www.forestry.gov.cn/main/65/content-554361.html

State Forestry Administration. (2000d). 关于开展2000年长江上游、黄河上中游地区退耕还林(草)试点示范工作的通知 [Notice of the pilot phase of the grain for green Program in 2000]. http://tghl.forestry.gov.cn/portal/tghl/s/2166/content-448772.html

State Forestry Administration. (2001a). 退耕还林技术模式 [Technical model of grain for green program]. Beijing: Forestry Press.

State Forestry Administration. (2001b). *China forestry development annual report*. Beijing: China Forestry Press.

State Forestry Administration. (2001c). 国家林业局关于印发退耕还林工程生态林与经济林认定标准的通知 [Definition of ecological forests and economic forest for the grain for green program]. http://www.forestry.gov.cn/portal/jjlxx/s/1960/content-163532.html

State Forestry Administration. (2002). *China forestry development annual report*. Beijing: China Forestry Press.
State Forestry Administration. (2003a). *Key forestry programs report*. Beijing: State Forestry Administration. Accessed online 28 Apr 2014 at http://www.forestry.gov.cn/CommonAction.do?dispatch=index&colid=67
State Forestry Administration. (2003b). *China forestry development annual report*. Beijing: China Forestry Press.
State Forestry Administration. (2003c). *Report for monitoring and assessment of the socio-economic impacts of China's key forestry programs*. Beijing: State Forestry Administration.
State Forestry Administration. (2003d). 中共中央 国务院关于加快林业发展的决定 [State council's decision on speeding up forestry development]. http://www.forestry.gov.cn/Zhuanti/content_gzhy/267626.html
State Forestry Administration. (2003e). 2002 年林业大事记 [The chronology of the forestry events in 2002]. http://www.forestry.gov.cn/main/60/content-100.html
State Forestry Administration. (2004a). *Key forestry programs report*. Beijing: State Forestry Administration. Accessed online 28 Apr 2014 at http://www.forestry.gov.cn/CommonAction.do?dispatch=index&colid=67
State Forestry Administration. (2004b). *China forestry development annual report*. Beijing: China Forestry Press.
State Forestry Administration. (2005a). *Key forestry programs report*. Beijing: State Forestry Administration. Accessed online 28 Apr 2014 at http://www.forestry.gov.cn/CommonAction.do?dispatch=index&colid=67
State Forestry Administration. (2005b). *China forestry development annual report*. Beijing: China Forestry Press.
State Forestry Administration. (2005c). The progress of the grain for green program. http://www.forestry.gov.cn/
State Forestry Administration. (2005d). *Report for monitoring and assessment of the socio-economic impacts of China's key forestry programs*. Beijing: State Forestry Administration.
State Forestry Administration. (2005e). 吉林省人民政府关于表彰全省天然林资源保护工程一期建设先进单位和先进个人的决定 [Acknowledgement of Pioneer counties and individuals in the natural forest protection program in Jilin Province]. http://www.jl.gov.cn/zwgk/gwgb/szfwj/jzh/201108/t20110804_1041764.html
State Forestry Administration. (2006a). *Key forestry programs report*. Beijing: State Forestry Administration. Accessed online 28 Apr 2014 at http://www.forestry.gov.cn/CommonAction.do?dispatch=index&colid=67
State Forestry Administration. (2006b). *China forestry development annual report*. Beijing: China Forestry Press.
State Forestry Administration. (2007a). *Key forestry programs report*. Beijing: State Forestry Administration. Accessed online 28 Apr 2014 at http://www.forestry.gov.cn/CommonAction.do?dispatch=index&colid=67
State Forestry Administration. (2007b). *China forestry development annual report*. Beijing: China Forestry Press.
State Forestry Administration. (2008a). *Key forestry programs report*. Beijing: State Forestry Administration. Accessed online 28 Apr 2014 at http://www.forestry.gov.cn/CommonAction.do?dispatch=index&colid=67
State Forestry Administration. (2008b). *China forestry development annual report*. Beijing: China Forestry Press.
State Forestry Administration. (2009). *China forestry development annual report*. Beijing: China Forestry Press.
State Forestry Administration. (2010). *China forestry development annual report*. Beijing: China Forestry Press.
State Forestry Administration. (2011). *China forestry development annual report*. Beijing: China Forestry Press.
State Forestry Administration. (2012a). *China forestry development annual report*. Beijing: China Forestry Press.

State Forestry Administration. (2012b). *Report for monitoring and assessment of the socio-economic impacts of China's key forestry programs*. Beijing: State Forestry Administration.

State Forestry Administration. (2012c). 权威解读 天然林资源保护工程二期政策 [Interpretation of the implementation plan for the second phase of the NFPP]. http://www.forestry.gov.cn//portal/main/s/72/content-480476.html

State Forestry Administration. (2013a). 机构简介 [Introduction of State Forestry Administration]. http://www.forestry.gov.cn/portal/main/xxgk/jgjj.html

State Forestry Administration. (2013b). *China forestry statistical analysis report*. Beijing: State Forestry Administration.

State Forestry Administration. (2013c). *Report for monitoring and assessment of the socio-economic impacts of China's key forestry programs*. Beijing: State Forestry Administration.

State Forestry Administration. (2000–2012). *China forestry development annual report*. Beijing: China Forestry Press.

State Forestry Administration. (2002–2011). *China forestry development annual report*. Beijing: China Forestry Press.

State Forestry Administration. (2003–2006). *China forestry development annual report*. Beijing: China Forestry Press.

State Forestry Administration. (2006–2008). *China forestry development annual report*. Beijing: China Forestry Press.

Statistics Bureau of Zhangye Prefecture. (2002–2005). *Zhangye prefecture statistical yearbook 2001–2004*, Zhangye Prefecture, Gansu Province. Unpublished.

Sun, C., & Liqiao, C. (2006). *A study of policies and legislation affecting payments for watershed services in China*. Beijing/London: Research Center of Ecological and Environmental Economics/International Institute for Environment and Development.

Tan, X. Y., & Chen, S. (1999). The current situation of food surpluses and countermeasures. *Problems of agricultural economy, 7*, 9–13 (in Chinese).

Tang, K. (2000). The harmonious development of reforestation and food security. *Journal of Water and Soil Reservation, 8*, 35–37 (in Chinese).

Tao, Z. (1958). Barge "Grain Production Limited Theory". *Red Flag, 8*.

Tao, Y. (1994). *Historical changes of China forest*. Beijing: China Forestry Publishing House.

Tao, R., & Qin, P. (2007). How has rural tax reform affected farmer and local governance in China? *China and World Economy, 15*(3), 19–32.

Tao, R., Xu, Z., & Xu, J. (2004). 退耕还林, 粮食政策与可持续发展 [Grain for green project, grain policy and sustainable development]. *Social Sciences in China, 6*, 25–38.

The Central People's Government of the People's Republic of China. (2013). 中华人民共和国行政区划 [Administrative divisions of the People's Republic of China]. http://www.gov.cn/test/2005-06/15/content_18253.htm

Trac, C. J., Harrell, S., Hinckley, T. M., & Henck, A. C. (2007). Reforestation programs in southwest China: Reported success, observed failure, and the reasons why. *Journal of Mountain Science, 4*(4), 275–292.

U.S. Central Intelligence Agency. (1986). *China agricultural regions*. Washington, DC (USA): U.S. Central Intelligence Agency.

Uchida, E., Xu, J., Xu, Z., & Rozelle, S. (2004). *Are the poor benefiting from China's conservation set-aside program?* (Working paper). Davis: University of California.

Uchida, E., Xu, J. T., & Rozelle, S. (2005). Grain for Green: Cost-effectiveness and sustainability of China's conservation set-aside program. *Land Economics, 81*(2), 247–264.

Uchida, E., Xu, J. T., Xu, Z. G., & Rozelle, S. (2007). Are the poor benefiting from China's land conservation program? *Environment and Development Economics, 12*(4), 593–620.

Uchida, E., Rozelle, S., & Xu, J. (2009). Conservation payments, liquidity constraints, and off-farm labor: Impact of the Grain-for-Green Program on rural households in China. *American Journal of Agricultural Economics, 91*(1), 70–86.

USDA. (2000). Land retirement. In U. S. Department of Agriculture, Economic Research Service Resource Economics Division (Ed.), *Agricultural resources and environmental indicators*. Washington, DC: U. S. Department of Agriculture, Economic Research Service.

Wan, C. (2011). Development of Chinese forestry industry: Government function improving and financial support system building. *Eastern Academic Forum*, 113–117. http://www.seiofbluemountain.com/upload/product/201103/2011dt04a1.pdf. Accessed 10 Jan 2014.

Wang, J. (1994). *Sichuan forestry (1)*. Chengdu: Sichuan Science and Technology Publishing House.

Wang, Z. (2000). *China's forestry ecological construction and China's national conditions* (China National Research Report, China National Committee Report Compiled). Beijing: The Central Literature Press.

Wang, X. Y. (2008). The rural reform and rural social changes. In Q. Li (Ed.), *Social changes in China, 1978–2008* (pp. 57–89). Beijing: Social Science Academic Press, Chapter 3 (in Chinese).

Wang, D. (2013, January 1). The reform of government authorities in towns and townships. English Edition of *Qiushi Journal, 5*(1). http://english.qstheory.cn/politics/201302/t20130227_213706.htm. Accessed 10 Jan 2014.

Wang, W., & Delang, C. O. (2011). Chinese forest policies in the age of ideology (1949–1977). *International Forestry Review, 13*(4), 416–443.

Wang, C., & Maclaren, V. (2011). Evaluation of economic and social impacts of the sloping land conversion program: A case study in Dunhua County, China. *Forest Policy and Economics, 14*(1), 50–57.

Wang, C. M., & Maclaren, V. (2012). Evaluation of economic and social impacts of the sloping land conversion program: A case study in Dunhua County, China. *Forest Policy and Economics, 14*(1), 50–57.

Wang, Y., Arrhenius, G., & Zhang, Y. (2001). Drought in the Yellow River – An environmental threat to the coastal zone. *Journal of Coastal Research Special Issue, 34*, 503–515.

Wang, G., Liu, G., & Zhou, S. (2003). 黄土高原土壤干层研究述评 [Research advance of soil dried layer on Loess Plateau]. *Journal of Soil Water Conservation, 6*, 156–159.

Wang, C., Ouyang, H., Shao, B., Maclaren, V., & Tian, Y. (2005, July 25–29). *Assessment of the sustainability of Grain for Green in Northeast China*. Proceedings of IGARSS '05, IEEE International, Geoscience and Remote Sensing Symposium, Seoul, Korea (Vol. 8, pp. 5695–5698).

Wang, G., Innes, J. L., Lei, J., Dai, S., & Wu, S. W. (2007a). China's forestry reforms. *Science, 318*(5856), 1556–1557.

Wang, H., Yan, C., & Jiang L. (2007b). 基于退耕还林工程的农民利益保障与增收的中西部比较研究——以江西省与宁夏自治区退耕还林工程为例 [Comparative study on interest security and income increase of farmers in the middle and western regions of China based on the restoring the reclaimed land to forest project – A case of the restoring the reclaimed land to forest project in Jiangxi Province and Ningxia autonomous region]. *Acta Agriculturae Universitatis Jiangxiensis, 2*, 318–322.

Wang, X., Lu, C., Fang, J., & Shen, Y. (2007c). Implications for development of grain-for-green policy based on cropland suitability evaluation in desertification-affected north China. *Land Use Policy, 24*(2), 417–424.

Wang, Z., Wang, X., Shi, Y., Pan, L., Yu, X., et al. (2007d). 三峡库区秭归县退耕还林工程水土保持效益研究 [Effects of soil and water conservation of the conversion of farmland to forest in Zigui County of the Three Gorges Reservoir Region]. *Journal of Soil and Water Conservation, 1*, 68–72.

Wenhua, L. (2004). Degradation and restoration of forest ecosystems in China. *Forest Ecology and Management, 201*(1), 33–41.

World Bank. (2001). *China air, land, and water: Environmental priorities for a new millennium*. Washington, DC: World Bank.

World Bank. (2014). *GINI index*. http://data.worldbank.org/indicator/SI.POV.GINI?page=2

World Wildlife Fund and State Forest Administration. (2003). *Case studies of China's grain-for-green Policy*. Beijing: Science Press.

Wu, S. R. (2004). Extension strategies in the Sloping Land Conversion Program in China: An analysis of their strengths and limitations. In *Proceedings of the 7th Extension Working Party Symposium "Communication Strategies for Multiple Partner Involvement in Forestry Extension"*, held 27 September–1 October 2004, Orvieto and Rome, pp. 45–59.

Wu, B., Zhang, Z., & Tang, L. (2007). Chapter 2: Forest rehabilitation in mainland China. In D. K. Lee (Ed.), *Keep Asia Green volume II "Northeast Asia"*. Vienna: International Union of Forest Research Organizations.

Wuqi GfG Office. (2007). *A summary report of implementation of the Sloping Land Conversion Program in Wuqi County*. Shaanxi Province: Wuqi GfG office.

Wuqi Statistics Bureau. (2006). *Statistics yearbook of Wuqi County 2005*. Wuqi County: Chongqing Municipality.

Xi, W., Bi, H., & He, B. (2012). Chapter 4: Forest landscape restoration in China. In W. Xi, H. Bi, & B. He (Eds.), *A goal-oriented approach to forest landscape restoration*. Heidelberg: Springer.

Xiao, D. (2008). *The history of People's Republic of China, volume 10. Turning point in history: Re-examination of the cultural revolution and the policy of reform and opening (1979–1981)*. Hong Kong: The Chinese university of Hong Kong Press (in Chinese).

Xiao, W., Dai, G., & Zhang, S. (2010, September 13–17). *China's strategy and financing for forestry sustainable development*. Paper presented at the first meeting of the ad hoc expert group on forest financing, Nairobi. UN Forum on Forests Secretariat. http://www.un.org/esa/forests/pdf/aheg/aheg1/China_case_study.pdf. Accessed 10 Jan 2014.

Xie, C., Zhao, J., Liang, D., Bennett, J., Zhang, L., et al. (2006). Livelihood impacts of the Conversion of Cropland to Forest and Grassland Program. *Journal of Environmental Planning and Management, 49*(4), 555–570.

Xu, J. (2011). China's new forests aren't as green as they seem. *Nature, 477*(7365), 371.

Xu, J., & Cao, Y. (2001). The socioeconomic impacts and sustainability of the SLCP. In J. T. Xu, et al. (Eds.), *Implementing the Natural Forest Protection Program and the Sloping Land Conversion Program: Recommendations, lessons and policy* (CCICED Task Force on Forests and Grasslands). Beijing: China Forestry Publishing.

Xu, J. T., & Cao, Y. Y. (2002). 退耕还林还草的可持续发展问题 [Sustainability of Green for Green Program) (in Chinese]. *International Economic Review, Z2*, 56–60.

Xu, J., Tao, R., & Xu, Z. (2003). *Sloping Land Conversion Program in West China – Analysis of household impact and sustainability* (Working Paper). Beijing: Centre for Chinese Agricultural Policy.

Xu, Z. G., Bennett, M. T., Tao, R., & Xu, J. T. (2004). China's Sloping Land Conversion Program four years on: Current situation, pending issues. *International Forestry Review, 6*(3–4), 317–326.

Xu, J., Yin, R., Li, Z., & Liu, C. (2006a). China's ecological rehabilitation: Unprecedented efforts, dramatic impacts, and requisite policies. *Ecological Economics, 57*(4), 595–607.

Xu, Z., Xu, J., Deng, X. Z., Huang, J. K., Uchida, E., et al. (2006b). Grain for Green versus grain: Conflict between food security and conservation set-aside in China. *World Development, 34*(1), 130–148.

Xu, J. T., Tao, R., & Xu, Z. G. (2006c). 退耕还林: 成本有效性, 结构调整效应与经济可持续性——基于西部三省农户调查的实证分析 [Sloping land conversion program: Cost-effectiveness, structural effect and economic sustainability]. *中国林业技术经济理论与实践* [Economical Theory and Practice of China's Forestry Technology], June 10–28.

Xu, J., Chen, L., Lu, Y., & Fu, B. (2007). Sustainability evaluation of the Grain for Green Project: From local people's responses to ecological effectiveness in Wolong Nature Reserve. *Environmental Management, 40*(1), 113–122.

Xu, J., Tao, R., Xu, Z., & Bennett, M. T. (2010). China's Sloping Land Conversion Program: Does expansion equal success? *Land Economics, 86*(2), 219–244.

Xue, Z. (2007). On difficulties and counter measures for constructing new socialist countryside in mountainous counties: A case study of Wuqi County as an example for Converting Farmland into Woodland. *Forestry Economics, 12*, 41–43 (in Chinese).

Yang, R. (1994). Cultivated land soil and water loss and control measures in China. *Bulletin of Soil and Water Conservation, 14*(2), 32–36.

Yang, X. D. (2005). Case study on benefits evaluation of Grain for Green Project. *Green China, 4*, 27–29.

Yang, S. (2006). 关于退耕还林"十一五"政策建议——川贵两省退耕还林调研思考 [Policy recommendations of Grain to Green Program during "Eleventh Five-Year Plan" – From experiences of Sichuan and Guizhou]. *Forestry Economics, 9*, 7–10.

Yang, G., Ding, G. D., Zhao, T. N., & Sun, B. P. (2006). Study on the effects of returning cropland to forest in loessy hilly region on soil and water conservation – Taking Wuqi County in Shaanxi Province as an example. *Bulletin of Soil and Water Conservation, 26*, 88–99.

Yao, S., & Li, H. (2010). Agricultural productivity changes induced by the sloping land conversion program: An analysis of Wuqi county in the Loess Plateau region. *Environmental Management, 45*(3), 541–550.

Yao, S., Guo, Y., & Huo, X. (2010). An empirical analysis of the effects of China's land conversion program on-farmers' income growth and labor transfer. *Environmental Management, 45*(3), 502–512.

Ye, Y. (2006). *China's agricultural cooperation movement research*. Beijing: Intellectual Property Press.

Yeh, E. T. (2009). Greening western China: A critical view. *Geoforum, 40*(5), 884–894.

Yi, F., Xu, J., & Xu, Z. (2006). 退耕还林经济影响再分析 [Reanalysis of the economic impact of the Grain for Green Program]. *Chinese Rural Economy, 10*, 28–36.

Yin, R. (2003). Chapter 4: Central characteristics of reform: Measures of the effects of improved property rights, a stable policy environment, and environmental protection.

Yin, R., & Liu, C. (2011). *The implementation and impacts of China's largest payment for Ecosystem Services Program as revealed by longitudinal household data*. http://www.indiana.edu/~workshop/papers/Yin_paper.pdf. Accessed 10 Jan 2014.

Yin, R., Liu, T., Yao, S., & Zhao, M. (2013). Designing and implementing payments for Ecosystem Services Programs: Lessons learned from China's cropland restoration experience. *Forest Policy and Economics, 35*, 66–72.

Yin, R. S, Xu, J. T., Li, Z., & Liu, C. (2005). *China's ecological rehabilitation: The unprecedented efforts and dramatic impacts of reforestation and slope protection in Western China* (China Environment Series 7, pp. 17–32). Washington, DC (USA): Woodrow Wilson International Center for Scholars

Yu, X. (1989). *The history of forest industry in Heilongjiang Province*. Haerbin: Heilongjiang Forestry Industry Bureau.

Yu, C. (1991). *Contemporary China's township enterprise*. Beijing: China Social Science Press.

Yu, B., & Lu, C. (2006). Change of cultivated land and its implications on food security in China. *Chinese Geographical Science, 16*(4), 299–305.

Yuan, D. S. (2010). Institutional innovation of Government Affairs Public. *Academic Journal of Zhongzhou, 180*, 16–20.

Yuan, Z. C., Gang, F., & Wing, T. W. (1996). Chapter 2: Chinese economic reforms: Past successes and future challenges. In T. W. Wing (Ed.), *Economies in transition: Comparing Asia and Eastern Europe*. Cambridge, MA: The MIT Press.

Zha, Y., & Gao, J. (1997). Characteristics of desertification and its rehabilitation in China. *Journal of Arid Environments, 37*(3), 419–432.

Zhang, H. (1989). Some comparisons in forest laws between China and Japan. *World Forestry Research, 1*, 64–68 (in Chinese).

Zhang, H. B. (1998). The evolution of China's environmental diplomacy. *World Economics and Politics, 11*, 38–44 (in Chinese).

Zhang, L. (2008). Chapter 9: Reform of the forest section in China. In P. Durst, C. Brown, J. Broadhead, R. Suzuki, R. Leslie, A. Inoguchi (Eds.), *Re-inventing forestry agencies: Experiences of institutional restructuring in Asia and the Pacific*. Rome: Food and Agriculture Organization of the United Nations. http://www.fao.org/docrep/010/ai412e/AI412E13.htm. Accessed 10 Jan 2014.

Zhang, Y. (2010). Application analysis of participatory approach in the grain for green project [J]. *Journal of Beijing Forestry University, 9*(1), 106–109 (in Chinese).

Zhang, W., & Liu, C. (2005). The impact of environmental policy on household income and activity choice: Evidence from Sandstorm Source Control Program in North China. In *American Agricultural Economics Association annual meeting*, Rhode Island.

Zhang, L., Luo, R., Liu, C., & Rozelle, S. (2006). Investing in rural China. *The Chinese Economy, 39*(4), 57–84.

Zhang, L., Bennett, J., Dai, G., Xie, C., Jia, Q., et al. (2008a). 中国退耕还林政策成本效益分析 [Cost-benefit analysis on the Program for Conversion of Cropland to Forest Land in China]. Beijing: Economic Science Press.

Zhang, L., Qin, T., & Mol, A. P. J. (2008b). Payment for environmental services: The Sloping Land Conversion Program in Ningxia autonomous region of China. *China and World Economy, 16*(2), 66–81.

Zhang, X. Y., Chen, J., & Zhao, Y. (2011). 退耕还林对陕北地区生态环境的影响 [Impact of China's Grain for Green Project on the ecological environment in Northern Shaanxi Province]. *Beijing Agriculture, 27*, 48–51.

Zhao, Y. T. (2010). Views on conversion of cropland into forest under current situation. *Research of Soil and Water Conservation, 17*(4), 276–278 (in Chinese).

Zhao, X., Lv, X., & Dai, J. (2010). Impact assessment of the "Grain for Green Project" and discussion on the development models in the mountain-gorge regions. *Frontiers of Earth Science in China, 4*(1), 105–116.

Zheng, Y., & Jiang, S. (2011). *Analysis of eutrophic nutrient sources and benefit of Grain for Green in Sancha Lake*. Paper presented at the international conference on Remote Sensing, Environment and Transportation Engineering (RSETE), 24–26 June 2011, Nanjing University of Science and Technology, China.

Zhi, L. (2004). Research on the multiple goal characteristics of our country's relief by grain from the backgrounds of Chinese and foreign returning farmland to forests. *Forest Economy, 7*, 29–32.

Zhou, H. (2001a). State council approves Six Key Forestry Programs. *People's Daily*.

Zhou, J. (2001b). A brief introduction to six major forestry programs in China [我国六大生态林业重点工程简介]. *Ecological Economy, 10*, 88.

Zhou, S., Yin, Y., Xu, W., Jia, Z., Caldwelld, I., et al. (2007). The costs and benefits of reforestation in Liping County, Guizhou Province, China. *Journal of Environmental Management, 85*(3), 722–735.

Zhou, D., Zhao, S., & Zhu, C. (2011). 退耕还林工程对黄土高原土地利用/覆被变化的影响——以陕西省安塞县为例 [Impacts of the Sloping Land Conversion Program on the land use/cover change in the loess plateau: A case study in Ansai county of Shaanxi province, China]. *Journal of Natural Resources, 26*, 1867–1878.

Zhu, L. (2008). Impacts of food and energy price hikes and proposed coping strategies. *China & World Economy, 16*(6), 35–45.

Zhu, Z., & Cheng, G. (1994). 中国土地沙质荒漠化 [China land desertification]. Beijing: Science Press.

Zong, Y., & Chen, X. (2000). The 1998 flood on the Yangtze, China. *Natural Hazards, 22*(2), 165–184.

Zuo, T., Zhou, S. K., & Zhong, B. F. (2003). 退耕还林工程政策实施过程分析 [The analysis of the implementation of grain for green program]. *Forestry Economics*, *4*, 20–22.

# Index




C.O. Delang, Z. Yuan, *China's Grain for Green Program*,
DOI 10.1007/978-3-319-11505-4

MIX
Papier aus verantwortungsvollen Quellen
Paper from responsible sources
FSC® C105338
FSC
www.fsc.org

If you have any concerns about our products,
you can contact us on
**ProductSafety@springernature.com**

In case Publisher is established outside the EU,
the EU authorized representative is:
**Springer Nature Customer Service Center GmbH**
**Europaplatz 3, 69115 Heidelberg, Germany**

Printed by Libri Plureos GmbH
in Hamburg, Germany